Time-Series Analysis and Cyclostratigraphy
Examining stratigraphic records of environmental cycles

Increasingly environmental scientists, palaeoceanographers and geologists are collecting quantitative records of environmental changes from sediments, ice cores, cave calcite, corals and trees. These records reveal climatic cycles lasting between one year and hundreds-of-thousands of years, and tidal cycles lasting from half a day to one and a half thousand years. The study of such records is known as cyclostratigraphy and the records themselves, time series. This book uses straightforward explanations of time-series analysis based on numerous original diagrams rather than formal mathematical derivations and equations.

All the main methods used in cyclostratigraphy are covered, including spectral analysis, cross-spectral analysis, filtering, complex demodulation, wavelet analysis and singular spectrum analysis. The problems of distortions of environmental signals during stratigraphic encoding are considered in detail, as are the practical problems of time-series analysis. Finally, there is a summary of the state of research into various types of tidal and climatic cycles and their cyclostratigraphic records. Extensive referencing allows ready access to the literature and the appendix provides a list of sources of computer algorithms.

This book provides the ideal reference for all those using time-series analysis to study the nature and history of sedimentary, climatic and tidal cycles. It is suitable for senior undergraduate and graduate courses in environmental science, palaeoceanography and geology.

GRAHAM WEEDON is Senior Lecturer in Geology at the University of Luton. His current research involves studying records of annual cycles in cave calcite, El Niño/Southern Oscillation (ENSO) cycles in varved sediment and Milankovitch cycles in ancient deep-sea sediments.

He received a D.Phil from Oxford University in 1987, and has participated in Ocean Drilling Program cruises, off Oman (1987), Brazil (1994) and New Zealand (1998). In 1999 he co-convened a Royal Society Meeting entitled 'Astronomical (Milankovitch) Calibration of the Geological Time Scale'.

T0205935

Time-Series Analysis and Cyclostratigraphy

Examining stratigraphic records of environmental cycles

Graham P. Weedon

CAMBRIDGE
UNIVERSITY PRESS

CAMBRIDGE UNIVERSITY PRESS
Cambridge, New York, Melbourne, Madrid, Cape Town, Singapore, São Paulo

Cambridge University Press
The Edinburgh Building, Cambridge CB2 2RU, UK

Published in the United States of America by Cambridge University Press, New York

www.cambridge.org
Information on this title: www.cambridge.org/9780521620017

© Cambridge University Press 2003

First published 2003
This digitally printed first paperback version 2005

A catalogue record for this publication is available from the British Library

Library of Congress Cataloguing in Publication data

Weedon, Graham P. (Graham Peter), 1962–   .
Times-series analysis and cyclostratigraphy: examining stratigraphic records
of environmental cycles/Graham P. Weedon.
        p.   cm.
Includes bibliographical references and index.
ISBN 0 521 62001 5
1. Cyclostratigraphy. 2. Time-series analysis. I. Title.
QE651.5. W44 2003
551.7′01′51955–dc21   2002067374

ISBN-13 978-0-521-62001-7 hardback
ISBN-10 0-521-62001-5 hardback

ISBN-13 978-0-521-01983-5 paperback
ISBN-10 0-521-01983-4 paperback

For Alexis and 'Felix-man'.

Contents

Preface

This book is designed to introduce the main methods used in the examination of quantitative records of ancient environmental changes. Such records are obtained from sources as diverse as the composition of sedimentary rocks, the varying percentages of microfossils and the thicknesses of growth bands in corals. These data sets, or time series, describe environmental changes lasting from half a day to millions of years. The emphasis of the book is on explaining concepts, procedures and problems *not* the details of the mathematics. I have avoided equations and derivations and have instead tried to employ simple diagrams in the explanations. This is because palaeoceanographers, environmental scientists, palaeoclimatologists, sedimentologists and palaeontologists sometimes find it easier to grasp new ideas graphically, rather than through formal mathematical treatments. There are, of course, many texts devoted to mathematical explanations, but this book attempts to explain time-series analysis to non-mathematicians in an accessible form.

Examination of ancient examples of varves and sedimentary cycles linked to orbital-climatic forcing (Milankovitch cycles, explained in Chapter 6) using time-series analysis began in the early 1960s. My own work spans Silurian to Recent cyclic sediments and includes the study of cores from three oceans with an emphasis on orbital-climatic (Milankovitch) forcing (Ocean Drilling Program Legs 117, 154 and 181). However, over the last few years the fastest growth in the use of time-series analysis has been amongst environmental scientists studying short period cyclicity related to phenomena such as El Niño and the Southern Oscillation and millennial-scale cycles and sedimentologists interested in stratigraphic records of tidal cycles. So despite my own perspectives, which have undoubtedly influenced the makeup of the book, I have tried to provide a treatment that is useful to all those interested in time series obtained from a stratigraphic context.

Throughout the book, in addition to demonstrations using artificial time series, I have used an example of a real cyclostratigraphic data set, obtained from British Lower Jurassic strata, to illustrate the major principles. Although this example is not ideal for every situation it does help to understand the procedures described as applied to real data. In several places I have made reference to issues concerned with the processing of sound and electronic digital signals in order to exemplify time series issues from everyday life. I have assumed that the reader is familiar with the concepts of standard deviation, correlation coefficients, moving averages, the normal and chi-squared distributions and covariance (e.g. Williams, 1984; Davis, 1986).

The subject of time-series analysis is full of jargon, so the first use of an important term is placed in **bold** along with the most common synonyms to allow easier reference to other publications. All the computations for the book illustrations used a modest PC running programs based on modifications of the published FORTRAN algorithms listed in the Appendix. Due to their central role, and at the risk of repeating the text, the figures have captions that allow them to almost 'stand alone'. To produce a consistent format all the figures are original and virtually all were created using the package *Microcal* Origin 6.

The intention has been to provide a text that will appeal to many disciplines while recognizing that some material may not appear directly relevant. This book is necessarily only an introduction, but if it helps to encourage new researchers into the field it will have served its purpose.

Graham P. Weedon
February 2002

Acknowledgements

This book took five years to write, but it has been great fun. A wide range of literature and mathematical techniques needed to be reviewed and associated computer algorithms assessed, and much of relevance was published during writing (one-third of the references cited date from 1997 or later). Fortunately many people have helped me both during and at the end of the writing.

I had useful discussions with, and suggestions from, Steve Clemens, Ian Hall, Tim Herbert, Linda Hinnov, Paul Olsen, Walther Schwarzacher and Nick Shackleton. Heiko Pälike helped with the wavelet spectrum in Fig. 4.19 and Paul Cole (the '2 m scale' in the photograph on the front cover) helped me with EPS formats for Figs. 1.8, 1.9 and 6.1. Additionally, Shaun Hines processed Fig. 1.9 electronically and Harvey Alison sorted out various difficulties with Origin files. Data for figures were generously supplied by Ben Walsworth-Bell (Fig. 2.5), Al Archer (Fig. 6.17) and Jerry McManus (Fig. 6.23). Thanks to Cambridge University Library and the Cambridge Department of Earth Sciences Library for allowing me access.

My thanks are due to Simon Crowhurst, Gavin Dunbar, Pat Fothergill, Kate Ravilious and Ben Walsworth-Bell, for reading early versions of the manuscript. Particular sections of the finished manuscript were kindly read by Al Archer, Gerard Bond, Angela Coe, Hugh Jenkyns and Jerry McManus. Individual chapters were read by Nick McCave, Heiko Pälike, Evelyn Polgreen, Walther Schwarzacher and Paul Shaw. Finally, the whole book was read by Simon Crowhurst and Ian Hall. My thanks to them all for their corrections and useful suggestions. Thanks also to Matt ('Write a book for me Graham') Lloyd and Sally Thomas at CUP for their support and to Sarah Price for copy-editing. Naturally, despite all this help I am responsible for any errors that may have escaped detection.

Chapter 1

Introduction

1.1   Cyclostratigraphic data

Increasingly, quantitative records of environmental change covering intervals of between half a day to millions of years are being sought by palaeoceanographers, environmental scientists, palaeoclimatologists, sedimentologists and palaeontologists. The 'media' from which these records are obtained range from sediments and sedimentary rocks to living organisms and fossils showing growth bands (especially trees, corals and molluscs), ice cores and cave calcite. This book is concerned with explaining the quantitative methods that can be employed to derive useful information from these records. Much of the discussion is concerned with explaining the problems and limitations of the procedures and with exploring some of the difficulties with interpretation. Most frequently environmental records are obtained from sedimentary sections making up the stratigraphic record and, using a rather broad definition, all the 'media' described above are 'stratigraphic'. The nature of cycles in environmental signals and in stratigraphic records are explored later. However, for now cycles can be thought of as essentially periodic, or regular, oscillations in some variable. The study of stratigraphic records of environmental cycles has been called cyclostratigraphy (Fischer *et al.*, 1990).

By regarding stratigraphic records of environmental change as signals, it is clear that the methods and interpretations reached during analysis must allow for the imperfections inherent in all recording procedures. In cyclostratigraphic data the environmental signal, which is 'encoded' during sedimentation, is often corrupted to some extent by interruptions caused by processes that are not part of the normal depositional system. Such processes, for sediments, include non-deposition, erosion, seafloor dissolution or event-bed deposition and they make the later recognition of the normal environmental signal more difficult. Yet the interruptions convey information themselves, and in some cases they result from the extremes of the normal environmental variations. For example, Dunbar *et al.* (1994), in their study of corals, pointed out that growth band thickness was related to sea surface temperature. However, episodes of unusually high sea surface temperatures cause growth band generation to stop completely for several years, thus interrupting the proxy temperature record.

As well as interruptions, the recording processes can introduce distortions that need to be taken into account. For example, accumulation rate variations and diagenesis frequently modify the final shapes of cyclostratigraphic data sets. In a similar manner to the interruptions, the distorting processes often depend on the nature of the environment. Hence, cyclostratigraphic data contain information about normal environmental variability, abnormal environmental variations and the processes that produce the records themselves. In other words, the stratigraphic information that is observed can be regarded as the product of many superimposed environmental and sedimentological, or metabolic, processes.

The methods described in this book are primarily concerned with detecting and describing regular cyclic environmental processes. Hence, the data are treated as though they consist of regular cycles plus irregular oscillations. The irregular components result from both normal and abnormal environmental conditions as well as the effects of sedimentation and diagenesis (or equivalent processes in skeletal growth, etc.). As explained below, there are sound geological reasons for using mathematics to search for regular cycles. Regular components of cyclostratigraphic data are often studied more easily than the irregular components. If methods could be developed to distinguish the various types and origins of the irregular components, much of value could be uncovered. Quantitative studies of the interruption and distortion processes will undoubtedly be useful for understanding ancient environmental and diagenetic mechanisms, but such investigations are relatively rare (e.g. Sadler, 1981; Ricken, 1986; Ricken and Eder, 1991; Ricken, 1993).

The idea that stratigraphic data consist of regular components – the signal, plus irregular components or noise – is based on a linear view of the processes involved. In reality non-linear processes abound in environmental systems (e.g. Le Treut and Ghil, 1983; Imbrie *et al.*, 1993a; Smith, 1994). In non-linear systems, the output does not vary in direct proportion to the input. There are many aspects of cyclostratigraphic data that cannot be easily investigated using the standard linear methods of analysis described in this book. From the perspective of non-linear dynamical systems, part of the irregular components can be considered to be as much a part of the environmental signal as the regular components (Stewart, 1990; Kantz and Schreiber, 1997). Some non-linear methods are described very briefly within Chapter 4 and some non-linear issues in signal distortion are considered in Chapter 5. Despite the view that non-linear approaches might explain more of the data than the linear methods, the latter are currently best understood mathematically and are the most frequently used.

A good demonstration of the success of the standard linear approach to cyclostratigraphic data concerns the time scale developed using late Neogene deep-sea sediments.

Hilgen (Hilgen and Langereis, 1989; Hilgen, 1991) and Shackleton *et al.* (1990) independently derived orbital cycle chronologies based on matching sedimentary cycles and oxygen isotope curves to the calculated history of insolation changes (Section 6.9). The results were at odds with the widely accepted radiometric ages that had been obtained using potassium-argon dating. Subsequently, improved radiometric dating and studies of sea-floor spreading rates confirmed the validity and utility of the so-called astronomical time scale approach (Wilson, 1993; Shackleton *et al.*, 1995a, 1999a). Consequently, a recent geochronometric scale for part of the Neogene has been based directly on orbital-cycle chronology rather than the traditional data derived from radiometrically calibrated rates of sea-floor spreading (Berggren *et al.*, 1995). In this case the standard, linear methods of time-series analysis have yielded results of fundamental importance to many other areas of the Earth Sciences.

1.2   Past studies of cyclic sediments

Examination of cyclic sediments intensified in the 1960s as modern depositional environments were better understood and conceptual models became more sophisticated. Historically sedimentologists were looking for explanations for cyclic stratigraphic sequences that did not simply require **random** (i.e. unconnected, meaning uncorrelated or 'independent') events. Perhaps if the underlying controls could be uncovered, more could be learnt about the environment of deposition. Cycle-generating processes were described as autocyclic if they originated inside the basin of deposition. Alternatively, allocyclic processes originated outside the basin (Beerbower, 1964). Coal measure cyclothems were a particular target for investigation since they had a wide range of interbedded lithologies, and resulted from a range of suspected autocyclic and allocyclic mechanisms. The definition of a cyclothem (Wanless and Weller, 1932) soon became contentious once the variety of lithological successions and inferred origins was appreciated (Duff *et al.*, 1967; Riegel, 1991). Simpler cyclic sections involving two alternating lithologies, often described as rhythmic, were often mentioned in reviews of cyclic sedimentation but, aside from sequences that were inferred to contain varves, they were little studied (e.g. Anderson and Koopmans, 1963; Schwarzacher, 1964).

In many early investigations, pattern recognition was centred on the analysis of the observed sequences of lithologies. This made sedimentological sense as the predictions of qualitative models could be compared with the observations. Of course no reasonably long stratigraphic section actually corresponded exactly to the pattern predicted by the models. Unfortunately, since it was easy to imagine situations where the expected or 'ideal cycle' (Pearn, 1964) was not encoded in the sedimentary rocks, it proved impossible to falsify the models. Duff and Walton (1962) argued that sedimentary cycles can be recognized as having a particular order of lithologies that frequently occur in a particular sequence. They called the most frequently occurring sequence a modal cycle. However, their definition of cyclicity was criticized as being so vague that it could include sequences that are indistinguishable from the result of random fluctuations – which would also exhibit modal cycles (Schwarzacher, 1975).

Markov chain analysis was used to test sequences for the presence of a **Markov property** or the dependence of successive observations (lithologies or numbers) on previous observations. This captured some of the concept of a 'pattern' in a cyclic sequence since it implied a certain preferred order to the observed lithologies. However, stratigraphic data as structured for Markov analysis apparently always have preferred lithological transitions, and thus never correspond to a truly independent random sequence (Schwarzacher, 1975). This is because environmental systems include a degree of 'inertia'. Even instantaneous changes in the 'boundary conditions' (e.g. sea level, rainfall, etc.) do not cause instantaneous changes in the environment. For example, it can be as much as a few years before the release, over a few weeks or months, of a large volume of sulphate aerosols into the atmosphere by a volcanic eruption causes a drop in global atmospheric temperatures (Stuiver et al., 1995). Therefore, the ubiquitous detection of a Markov property in cyclic sections merely indicated that there is a degree of 'smoothness' in the transitions between successive observations. Since virtually all physical systems exhibit inertia, the detection of a Markov property proved to be of little use for characterizing sedimentary cyclicity. Nevertheless, Markov analysis is useful when, for example, the particular order of lithologies helps in the description of sedimentological processes (e.g. Wilkinson et al., 1997).

Schwarzacher's (1975) book represented a landmark in the examination of sedimentary cyclicity. Instead of just examining the transitions between lithologies at bed boundaries in Markov chain analysis, he reasoned that the thickness of successive beds provided information of fundamental importance in the assessment of sedimentary cycles. This meant that the stratigraphic data should be collected as **time series**. Time series include any sequence of measurements or observations collected in a particular order. Usually the measurements are made at constant intervals of some scale of measurement such as cumulative rock thickness, geographic distance, time, growth band number, etc. Some authors have referred to data collected relative to a depth or thickness scale as 'depth series', but time series is actually the correct mathematical term for historical reasons (Schwarzacher, 1975; Priestley, 1981; Schwarzacher, 1993). The variable that is recorded need not be restricted to lithology of course, and this significantly widens the scope of potential investigations of sedimentary cyclicity. The quantitative techniques used for the study of such data are described as methods of **time-series analysis**.

Schwarzacher argued that to be meaningful the term 'sedimentary cycles' must refer to oscillations having perfectly or nearly perfectly constant **wavelength**. Only if the wavelength can be measured in time does one refer to the cycle's **period**. However, whether a time or thickness scale is being used, oscillations of constant wavelength are described by mathematicians as **periodic**, and those of nearly constant wavelength as **quasi-periodic**. Periodic or quasi-periodic cyclostratigraphic sections have repetitions of a particular observation (such as a particular rock type) at essentially constant stratigraphic intervals. To many mathematicians stratigraphic sections that do not exhibit this type of regularity should not be termed cyclic at all (Schwarzacher, 1975). Yet sedimentary cyclicity is a perfectly useful field term for sections with interbedded rock types where event deposition is not involved (Einsele *et al.*, 1991). The mathematician's approach would require mathematical investigations before the term sedimentary cyclicity could be applied. I argue here that 'cyclicity' and 'sedimentary cycles' are liable to be used by sedimentologists, however vaguely, for the foreseeable future. Instead I have used the terms **regular cycles** and **regular cyclicity** to denote oscillations in stratigraphic records that can be shown, using time-series analysis, to have near-constant wavelengths (i.e. rock thickness) or periods. The issue of nomenclature of cyclic sediments is currently being assessed by the Working Group on Cyclostratigraphy appointed by the International Subcommission on Stratigraphic Classification (Hilgen *et al.*, 2001).

In the late 1970s and 1980s two revolutions in sedimentological thinking profoundly influenced the study of cyclic sediments. Firstly, following extensive deep-sea drilling, improvements in the measurement of remnant magnetization and in radiometric dating, it became clear that the orbital or Milankovitch Theory of climatic change (Section 6.9) should be taken seriously as an explanation for the Pleistocene climate changes (Hays *et al.*, 1976; Imbrie and Imbrie, 1979; Imbrie *et al.*, 1984). This promoted intense interest in evidence for pre-Pleistocene orbital-climatic cycles (Sections 6.9.3 and 6.9.4). In the absence of accurate time scales, the most convincing demonstrations of ancient orbital-climatic cycles came from the time-series analysis methods advocated by Schwarzacher (1975) and used extensively by the palaeoceanographers examining Pleistocene sediments (Weedon, 1993). Pioneering time-series analyses of cyclic sequences (Preston and Henderson, 1964; Schwarzacher, 1964; Carrs and Neidell, 1966; Dunn, 1974) seem to have lacked the long data sets and time control needed to make sufficiently convincing cases for Milankovitch cyclicity to the wider community. Concurrent with the increased interest in Milankovitch cyclicity, the attempt to detect regular climatic and weather cycles possessing much shorter periods met with increasing success (Burroughs, 1992).

Meanwhile Vail *et al.* (1977, 1991) changed the way sedimentologists interpreted lithostratigraphic successions. By employing sequence stratigraphic methods, sedimentary sections can be divided into genetically related stratigraphic units. Stacks of sequences were explained in terms of changing base level, especially relative sea level. However, because a large variety of processes were believed to be ultimately responsible for sequence generation, a classification scheme based on the duration of sea level cycles was adopted (e.g. Vail *et al.*, 1991). This ranged from 'first order' sequences lasting more than 50 million years to 'sixth order' sequences formed in 10,000 to 30,000 years. Although the duration or 'order' of sequences was believed to provide a clue to their likely origin, regularity was not implied by their use of the term 'sea level cycle'. Nevertheless, the higher order sea level cycles were explained in terms of Milankovitch cycles, especially acting through glacio-eustasy (Goldhammer *et al.*, 1990; Naish and Kamp, 1997). The resulting sequences were termed parasequences if relatively complete, or simple sequences if bounded by stratigraphic gaps.

By the 1990s a more descriptive approach to cyclic sequences was being advocated (Einsele *et al.*, 1991). Studies of the links between ancient climatic changes and cyclic sedimentation increased, utilizing several Pleistocene models that include and exclude ice sheets and glacio-eustasy. The description and study of cyclic sedimentary sections became known as cyclostratigraphy (Fischer *et al.*, 1990). As discussed earlier, it is likely that future studies of irregular processes, particularly utilizing non-linear dynamic systems methods (Sections 4.7 and 4.8), will be fruitful. Studies of Milankovitch cyclicity are currently particularly concerned with the development of time scales based on counts of Milankovitch cycles for pre-Cenozoic sequences and matches with orbital 'templates' for the younger part of the Cenozoic (Section 6.9.3, Shackleton *et al.*, 1999a). However, a great deal of work on cyclostratigraphic signals is now being undertaken by palaeoceanographers, climatologists and environmental scientists concerned with climatic oscillations that have periods shorter than the orbital cycles (i.e. <20,000 years) as discussed in Chapter 6.

1.3   Time-series analysis – an introduction

As shown in Fig. 1.1 a simple oscillation can be described in terms of its **amplitude** and wavelength. Additionally, the position within the oscillation or its **phase angle** or **phase** (ranging from 0 to 360° or from 0 to 2π radians) can be measured from some sort of origin along the time or cumulative thickness/depth axis. Geologically the position of the origin is determined arbitrarily by wherever the data collection started. However, mathematically this type of simple oscillation is usually described using a sinusoid; if it starts at the mid-point of an oscillation it is a sine wave and if it starts at a maximum it is a cosine wave (Fig. 1.1). Sine and cosine waves are convenient for describing oscillations mathematically. To produce a sinusoid that starts at a phase angle of 45° it is only necessary to add together a sine and cosine wave of the same wavelength and the same amplitude (Fig. 1.2a). Any other starting angle can be generated by controlling the relative amplitudes of the sine and cosine waves used (Fig. 1.2b). Observational time series rarely have oscillations of such a simple shape, but more complicated shapes, such as cuspate waves with narrow troughs and long peaks, can be represented by adding sine and cosine waves with particular wavelengths (Section 5.2.4).

Observational time series are of course usually composed of many different wave-length oscillations. According to **Fourier's theorem**, any time series, no matter what shape it is provided it has some oscillations and no infinite values, can be recreated by adding together regular sine and cosine waves having the correct wavelengths and amplitudes. Sine and cosine waves form a set of so-called **orthogonal functions**. Or-thogonal functions are simply groups of waves that can be added together to describe any time series, but none of the individual component waves can be constructed from combinations of other waves in the group. There are other sets of orthogonal functions, which can be used in place of sines and cosines (e.g. **Walsh functions**, Section 3.4.6,

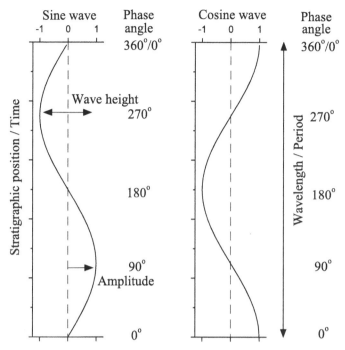

Fig. 1.1 The principal parameters needed to describe sinusoidal waves. Amplitude is measured as the maximum deviation from the zero line. Period (time interval) or wavelength (thickness interval) is defined as the interval from peak to peak or trough to trough, etc. The phase angle indicates the relative position within the complete cycle and is measured from the base of the data set. The phase angle, or more simply phase, ranges from 0 to 360 degrees (or 0 to 2π radians). Sine and cosine waves of the same wavelength are identical except that the phase differs by $90°$. The sine and cosine waves shown have a wavelength or period equal to the length of the whole time series.

Many cyclostratigraphic records from cores are labelled using depth or age from the top of the data downwards. However, it is standard lithostratigraphic practice, when studying sections exposed on land, to denote stratigraphic position by height or time increasing from the base upwards. Throughout this book the measurements or observations from the youngest strata are always located at the top of the time series plots (i.e. the top measurements relate to minimum depth or maximum height).

Beauchamp, 1984). However, most stratigraphic time series consist of approximately sinusoidal oscillations, so usually sine and cosine waves are the most naturally employed. Examination of time series using sines and cosines is often referred to as **Fourier analysis**.

Clearly it would be convenient to be able to take a time series and quickly assess how many regular component oscillations are present. This is most readily achieved by using **power-spectral analysis** (Chapter 3). Put simply the **power spectrum** shows the relative amplitudes (strictly squared amplitudes) and wavelengths or periods of all the regular components in the time series. By convention the horizontal axis of a power spectrum is plotted as **frequency** (frequency = 1/period) with highest frequencies (shortest oscillations) appearing on the right. Zero frequency refers to oscillations that

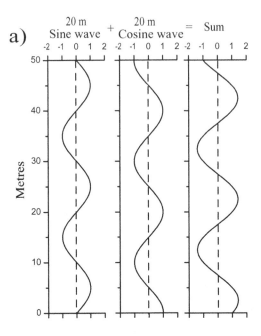

a)

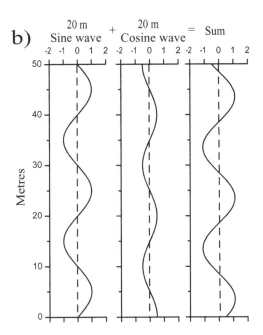

b)

Fig. 1.2 (a) When sine and cosine waves with the same wavelength and equal amplitude are added together, the resulting sinusoid has a phase which is intermediate between that of the components (i.e. it differs by 45°). (b) Adding a sine wave with an amplitude of one unit to a cosine wave with an amplitude of half a unit produces a sinusoid with a phase of 67.5°. This means that any sinusoid can be considered to represent the sum of one sine and one cosine wave having the same wavelength and the correct relative amplitudes.

have wavelengths or periods exceeding the length of the whole data set. If the data are collected as a function of time, frequency is measured in 'numbers of cycles per time unit' which is usually shortened to 'cycles per time unit' (e.g. cycles per thousand years). If a thickness or depth scale is used then instead of frequency some authors refer to the **wave number** (i.e. wave number = 1/wavelength in thickness). However, for clarity in cyclostratigraphic studies it is usual for one to speak of frequency even though the units do not include a time element (e.g. cycles per metre).

Spectral analysis requires amplitude measurements determined as positive or negative deviations from some zero line. However, although the zero line is sometimes defined using the average or mean of the data, usually a more complicated definition is involved as discussed later (Section 3.2). Consequently, it is often confusing, when inspecting a time series plot, if the zero line is indicated, so this has only been illustrated for the time series in Figs. 1.1 and 1.2. Geologists often prefer to plot stratigraphic position or time running vertically up the page. However, frequently palaeoceanographers and environmental scientists plot data relative to time so that the time axis runs horizontally, with younger data on the left of the page. To simplify the layout of the figures I have plotted all the time series the same way so that either stratigraphic position or time runs up the page, hence the youngest data are found at the top.

The vertical axis of the spectrum is usually plotted as squared amplitude and by analogy with physics it is described as 'power' (energy per time interval), hence the name power spectrum. Since amplitude refers to deviation from the zero line, squared amplitude can be thought of as squared deviation and so sometimes one speaks of the **variance spectrum** (variance equals squared standard deviation). Occasionally amplitude, rather than squared amplitude, is plotted against frequency, so creating an **amplitude spectrum** (also known as a **magnitude spectrum**). If small spectral peaks need to be studied together with large peaks then the log of power is plotted against frequency. In electronic signal processing, for comparing power values the **decibel scale** is used (i.e. $10 \times \log_{10}$power) so that a power value of 0.01 equals -20 dB. (Note that for comparing voltages, analogous to amplitude in time-series analysis, decibels are calculated as $20 \times \log_{10}$voltage/amplitude.)

It is sometimes useful to be able to think of spectral analysis using physical analogies. Thus the rainbow effect produced by a glass prism acting on a beam of white light is a classical example of a spectrum. The brightness of different parts of the rainbow corresponds to the power and the various colours the frequency. The ear and brain similarly apparently analyse sound (fluctuating air pressure) as though it is a time series made up of components with different amplitude/power (loudness) and frequency (pitch, Taylor, 1965, 1976). Thus different parts of the brain are activated by different frequencies, though the size of the response depends on musical training/skill and the type of sound (e.g. Pantev et al., 1998).

Figure 1.3 illustrates an example where a 10 m sine wave has been added to a 2.78 m sine wave. The resulting time series, shown as 'Sum' on the right, would have looked different if the relative phase of the two components and/or the relative amplitudes

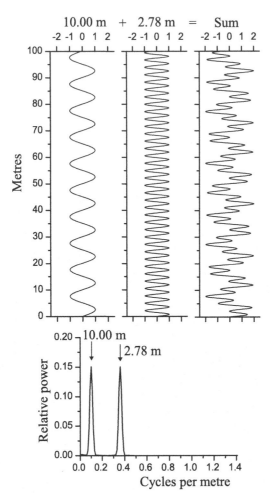

Fig. 1.3 Adding sinusoids with different wavelengths produces a time series with multiple frequency components. Power spectra are used to: (a) identify which frequency components are present (frequency = 1/wavelength); and (b) determine their relative amplitudes. In this case two sine waves of equal amplitude, but different wavelengths, have been added to produce the time series labelled "Sum". The corresponding power spectrum has peaks that occur at frequencies corresponding to the component wavelengths. The peaks are equal in height because the components have the same amplitude. Note that it is impossible to tell from the spectrum whether the components are sine or cosine waves – in other words the spectrum is independent of the phase of the components.

differed. The power spectrum in Fig. 1.3 shows that the time series consists of just two frequency components. When power spectra are generated all the phase information is discarded. As a result, changing the relative positions or phases of the 2.78 m oscillations relative to the 10.0 m oscillations, and hence the shape of the time series, would not influence the shape of the spectrum. The heights of the two spectral peaks in Fig. 1.3 are identical because the amplitudes of the component oscillations are identical. The larger the spectral peak, the greater the amplitude of the corresponding wavelength of oscillation and the greater its 'importance' in controlling the overall shape of the time series. The frequency of the spectral peaks can be read from the horizontal axis and indicates, of course, that oscillations with wavelengths of 10.00 m and 2.78 m are present in the time series.

Sinusoids with varying amplitude are said to exhibit **amplitude modulation** or **AM**. There are two types of amplitude modulation. In **heterodyne AM** the addition of two sinusoids with similar wavelengths creates a new single oscillation (Olsen, 1977). The new oscillation has a frequency that is the average of the frequencies of

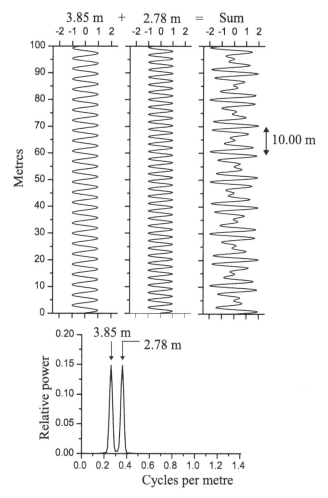

Fig. 1.4 Heterodyne amplitude modulation occurs when two sinusoids with similar frequencies are added together. Here sine waves with wavelengths of 3.85 m and 2.78 m are summed. The result is an oscillation with a wavelength of 3.29 m and a beat wavelength of 10 m. Note that the spectrum does not have a spectral peak corresponding to the beat frequency (0.1 cycles per metre) because there are no 10 m oscillations, just 10 m variations in amplitude. If the spectrum of this type of time series had a lower frequency resolution than illustrated here, the two spectral peaks would appear as one broad peak (Section 3.3.2).

the two added sinusoids. The variation in amplitude of the new oscillation is called the **beat** and this has a frequency that equals the difference in the frequencies of the added sinusoids (Taylor, 1965). For example, in Fig. 1.4 the addition of oscillations with frequencies of 1/3.85 m and 1/2.78 m generates an oscillation with a frequency of 1/3.29 m (i.e. = (1/2.78 m + 1/3.85 m)/2) and a beat with a wavelength of 10.00 m (i.e. 1/10.00 m = 1/2.78 m − 1/3.85 m). Note that the spectrum reveals the presence of the two original cycles, but no peak at the frequency of the beat frequency. This is because there are no oscillations with a 10 m wavelength in the record, just 10 m variations in amplitude.

Imposed AM occurs when a large period/wavelength signal (the beat) is used to vary the amplitude of another oscillation (the primary cycle, Olsen, 1977). In such cases, a peak at the primary cycle frequency dominates the spectrum, but small **combination tone** peaks are generated on either side due to the imposed beat frequency (discussed further in Section 5.2.4). The frequencies of the combination tone peaks

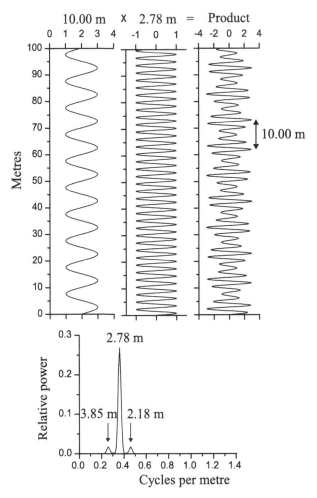

Fig. 1.5 Imposed amplitude modulation (AM) occurs when a separate signal is used to vary the amplitude of a regular sinusoid. In this case the amplitude of a 2.78 m cycle is made to vary between 1.0 and 1.5 by multiplication by a cycle with a wavelength of 10 m. The resulting time series looks very similar to that in Fig. 1.4, but the spectrum bears the hallmark of imposed AM with combination tone peaks on either side of the primary frequency. As for heterodyne AM there are no 10 m oscillations, so the spectrum does not contain a spectral peak at 0.1 cycles per metre.

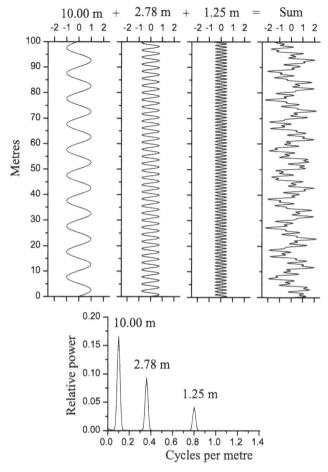

Fig. 1.6 When three sine waves of different wavelengths are added together, the resulting time series can look quite complicated. Nevertheless, as power spectra are unaffected by phase, the three component wavelengths and their relative amplitudes are readily determined.

correspond to the primary oscillation frequency minus the beat frequency and to the primary oscillation frequency plus the beat frequency (Taylor, 1965). In the example shown in Fig. 1.5, 10 m imposed AM of the 2.78 m oscillations generates sidebands at frequencies of 1/3.85 m (i.e. = 1/2.78 m – 1/10.0 m) and 1/2.18 m (i.e. = 1/2.78 m + 1/10.0 m). As for heterodyne AM, the spectrum of an imposed AM signal does not have a peak at the beat frequency because there are no 10 m oscillations present (just 10 m variations in amplitude). A crucial point is that power spectra reveal average power. Thus, except in rare clear-cut cases (e.g. Figs. 1.4 and 1.5), power spectra cannot be used to infer how the amplitude of an oscillation varies along the length of a time series.

Time series can look exceedingly complicated when only a few regular cycles are added together. In Fig. 1.6 the addition of three oscillations with different amplitudes results in a moderately complicated looking data set. The spectrum contains three

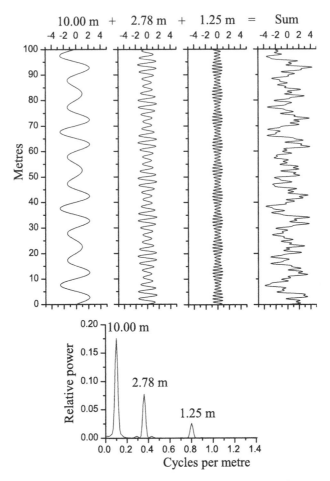

Fig. 1.7 Adding three sine waves with different wavelengths and varying amplitudes produces a time series that looks extremely complicated. It is unlikely that mere visual inspection of the summed time series illustrated would allow one to recognize that just three frequency components are present, or to determine the wavelengths involved. However, as the peak height depends only on (squared) *average* amplitude, the spectrum is very similar to the spectrum in Fig. 1.6.

spectral peaks, the relative peak heights indicating the average relative squared amplitudes of the component cycles. If the same three regular cycles have varying amplitudes along the series, the result looks considerably more complex (Fig. 1.7), but the corresponding spectrum is dominated by the same three peaks. Most people would be hard-pressed to recognize the presence of just three regular components in the time series of Fig. 1.7 merely by visual inspection. It would also be very difficult to establish the wavelengths involved. Therefore, except with very simple or especially characteristic data sets, it is unwise to claim the detection of regular cyclicity in a time series by visual inspection alone or by using simple analysis of the wavelength distribution of the oscillations (e.g. histograms of bed thickness).

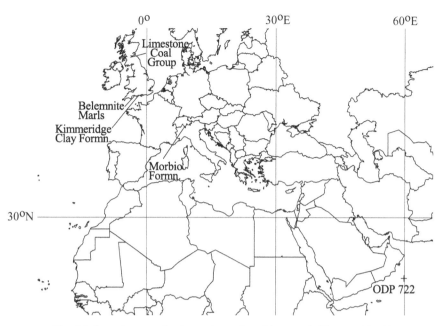

Fig. 1.8 Location map for the cyclostratigraphic records illustrated in Chapters 1 to 5. Formn. denotes formation.

Observational time series usually consist of the addition of tens or hundreds of regular sine and cosine components. This means that every possible regular frequency component has a non-zero amplitude. Consequently, spectral analysis of stratigraphic time series is used to look for spectral peaks that emerge from a background of spectral values.

This is an appropriate point to introduce the data sets used to illustrate this book. They are listed, according to the age of the strata from which they were obtained, in Table 1.1. Figures 1.8 and 6.1 show where these cyclostratigraphic records were obtained. Many of these records are based on oxygen-isotope records (see Faure, 1986 for an introduction to the determination and uses of $\delta^{18}O$ and an explanation of the delta notation). One of the records has been selected to help illustrate the various time series methods described in the book. The information comes from the Early Jurassic hemipelagic formation called the Belemnite Marls, which is approximately 190 million years old and exposed on the coast of Dorset, England (Fig. 1.8, Table 1.1, Weedon and Jenkyns, 1990, 1999). In the field much of this unit consists of interbedded light-grey marls, dark-grey marls and brown-black laminated shales. These lithologies form decimetre-scale bedding couplets that are grouped into metre-scale bundles (Fig. 1.9). In several intervals at the base of the formation, the bedding is barely visible. Towards the top the couplets and bundles are noticeably thinner (Fig. 1.9).

As for all stratigraphic records obtained from the Jurassic system, the absolute dating uncertainties create difficulties when assessing the duration of processes lasting less

Table 1.1. *Cyclostratigraphic time series used to illustrate the book.*

Period (time interval)	Borehole site or formation, location	Number of points	Sample interval	Variable	Time series description	Chapter
Recent (1840–1994)	Maiana Atoll, W. Pacific Ocean	928	0.166 years	$\delta^{18}O_{CORAL}$	Urban et al., 2000	6
Recent (1936–1982)	Galapagos Islands, E. Pacific Ocean	183	0.5 years	Ba/Ca_{CORAL}	Shen et al., 1992	6
Recent (1967–1711)	GISP2, Greenland	2047	0.125 years	$\delta^{18}O_{ICE}$	Section 6.2	6
Recent (1987–818)	GISP2, Greenland	1170	1.0 year	$\delta^{18}O_{ICE}$	Section 6.2	6
Recent (0.0–10.0 ka BP)	GISP2, Greenland	503	20 years	$\delta^{18}O_{ICE}$	Section 6.2	6
Late Pleistocene–Recent (0–80.8 ka BP)	GISP2, Greenland	405	200 years	$\delta^{18}O_{ICE}$	Section 6.2	6
Late Pleistocene–Recent (0–330 ka BP)	ODP 980, North Atlantic	928	Variable	$\delta^{18}O_{PF}$	McManus et al., 1999	6
Late Pleistocene–Recent (0–370 ka BP)	ODP 722, N.W. Indian Ocean	147	Variable	Ba/Al, Ti/Al	Weedon and Shimmield, 1991	5
Late Miocene–Recent (0–6 Ma BP)	ODP 677 and ODP 846, E. Pacific Ocean	2001	3000 years	$\delta^{18}O_{BF}$	Shackleton et al., 1990, 1995b	6
Early Miocene	Marine Molasse, Auribeau, S. France	154	Bundle thickness	Tidal bundle thickness	Archer, 1996	6
Late Jurassic	Kimmeridge Clay Formation, S. England	360	0.05 m	Magnetic susceptibility	Morgans-Bell et al., 2001; Weedon et al., 1999	5
Late Jurassic	Kimmeridge Clay Formation, S. England	117	0.1524 m	Photoelectric factor	Gallois, 2000; Morgans-Bell et al., 2001	5
Late Jurassic	Kimmeridge Clay Formation, S. England	90	0.2 m	%TOC	Morgans-Bell et al., 2001; Weedon et al., 1999	5
Early Jurassic	Morbio Formation, S. Switzerland	1024	0.01 m	Rock type code	Weedon, 1989	3
Early Jurassic	Belemnite Marls, S. England	798	0.03 m	$%CaCO_3$, %TOC	Weedon and Jenkyns, 1999	1–6
Late Carboniferous	Abbott Formation, S. Illinois, USA	208	Bed thickness	Tidal bed thickness	Archer, 1996	6
Early Carboniferous	Limestone Coal Group, C. Scotland	5236	0.01 m	Rock type code	Weedon and Read, 1995	5

Abbreviations: $\delta^{18}O_{PF}$, $\delta^{18}O$ in planktonic foraminifera; $\delta^{18}O_{BF}$, $\delta^{18}O$ in benthic foraminifera; ODP, Ocean Drilling Program; %TOC = percentage total organic carbon; ka, thousands of years; Ma, millions of years; BP, before present. Figures 1.8 and 6.1 provide location maps for these records.

Fig. 1.9 Photograph of the Belemnite Marls as exposed below Stonebarrow near Charmouth in Dorset, England. The whole formation is close to 24 m thick. The majority of the formation consists of beds of light-grey marl alternating with dark-grey marl and brown-black laminated shales. The alternations form couplets that are grouped into bundles. Both couplets and bundles become much thinner towards the top of the formation.

than about 10 million years (Gradstein *et al.*, 1994). Nevertheless, it is clear from time-series analysis that the decimetre-scale bedding couplets relate to the 20,000 orbital-precession cycle (Weedon and Jenkyns, 1999). Samples were collected throughout the Belemnite Marls at fixed 3-cm intervals and analysed for weight percent calcium carbonate and total organic carbon (or TOC, Fig. 1.10). Weedon and Jenkyns (1999) give instructions for obtaining a listing of these data. The results show that the light-grey marls have higher carbonate contents and less organic carbon than the dark-grey marls and laminated shales. Additionally, in the visually almost homogeneous interval towards the base (bed 110), there are the same types of compositional variations found elsewhere.

In Fig. 1.11 part of the time series of weight percent $CaCO_3$ from the Belemnite Marls is illustrated with the corresponding power spectrum. Although a large number of frequency components appear, the spectrum of the Belemnite Marls carbonate contents is clearly dominated by three main spectral peaks which emerge from the spectral background. These large spectral peaks are described as relating to regular sedimentary cycles, even though mathematically all the spectral values relate to regular components. Thus it is the geological interpretation of the spectrum that leads to the data being regarded as composed of a three-component regular cyclic 'signal' plus the irregular 'noise' accounting for the spectral background. The spectral peak

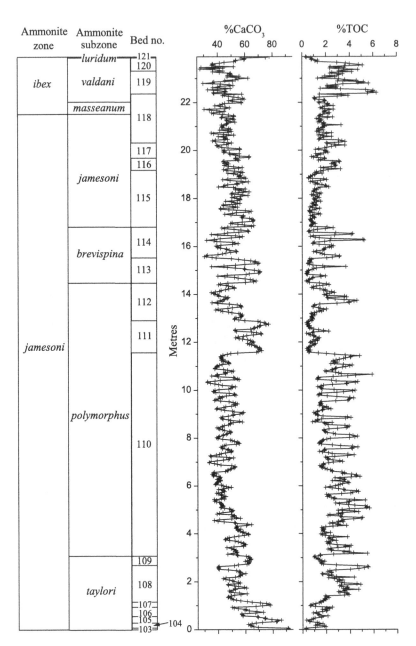

Fig. 1.10 Calcium carbonate (CaCO₃) and total organic carbon (TOC) time series from the whole of the Belemnite Marls. The top and base of the formation are marked by thin early diagenetic limestones that are associated with stratigraphic gaps (Weedon and Jenkyns, 1999). The formation covers the first two ammonite zones of the Pliensbachian Stage (Lower Jurassic). Bed numbers follow Lang *et al.* (1928). Note the persistence of couplet-like carbonate and organic carbon oscillations throughout bed 110, even though in the field the bedding is barely visible (Fig. 1.9).

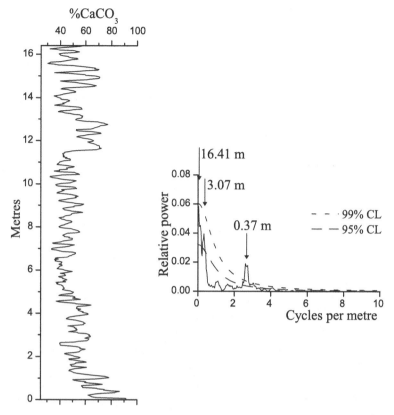

Fig. 1.11 A time series of calcium carbonate contents from the lower two-thirds of the Belemnite Marls. In real data there is some variation in the measured variable at every scale or wavelength. In general, lower frequency variations have a larger average amplitude than higher frequency variations. This produces a sloping continuum in the spectrum that rises towards the lowest frequency end. In the Belemnite Marls there are three main scales of variation in carbonate contents (Weedon and Jenkyns, 1999) that account for the labelled spectral peaks emerging from the sloping background continuum. The dashed lines indicate confidence levels (CL) and are used to distinguish peaks from the spectral background (Section 3.5).

For completeness the following describes the methods used to generate this spectrum (see Chapter 3 for explanations): (a) linear detrending of 548 data points, (b) zero-padding to 1024 points, (c) multi-taper spectral estimation using six data tapers. The 95% and 99% confidence levels (e.g. 99% CL) are used to identify those spectral peaks that cannot be attributed to the background noise.

labelled with a wavelength of 0.37 m relates to the bedding couplets and the peak labelled 3.07 m relates to the metre-scale bundles of couplets. In this case the noise in the time series can be regarded as the product of irregular oscillations in the environment plus measurement errors, which together partly explain the continuous spectral background.

Details of the methods needed to generate power spectra are given in Chapter 3. Many other methods of time-series analysis are available and some of the standard procedures are described in Chapter 4. The three most important methods are used to: (a) isolate regular cycles from the time series (filtering); (b) examine variations in cycle amplitude (amplitude demodulation); and (c) study the phase and amplitude relationships of pairs of variables obtained simultaneously from the same samples or time intervals (i.e. cross-spectral analysis). Chapter 5 is concerned with distortions of these environmental signals as recorded stratigraphically, as well as practical considerations in conducting time-series analyses of real data. Chapter 6 discusses the many environmental origins for the regular cycles that have been observed in stratigraphic records (from tidal to Milankovitch cycles).

In the context of cyclostratigraphic data sets, spectral analysis is usually the first procedure used, because it allows the detection of regular cyclicity and determination of wavelengths and average amplitudes. However, before the difficulties of generating and interpreting power spectra can be considered, it is essential that one is aware of the many issues concerning the construction of time series (Chapter 2).

1.4   Chapter overview

- Time-series analysis provides procedures for examining quantitative records of environmental variability. It was widely adopted in cyclostratigraphic studies following the vindication of the orbital-climatic theory (Milankovitch theory) in the early 1980s.

- Spectral analysis allows the detection of multiple regular cycles in a time series. Each regular component is characterized in terms of its frequency (= 1/period or 1/wavelength) and *average* power (= squared average amplitude).

- Regular amplitude modulation in a time series does *not* generate a spectral peak at the modulation frequency.

Chapter 2

Constructing time series in cyclostratigraphy

2.1   Introduction

This chapter is concerned with the acquisition of data and the generation of time series. The procedures discussed later in the book are based on the assumption that the time series used for analysis have been obtained with due regard for the issues addressed here. It is also assumed that care has been taken to ensure that the measurements or observations made are sufficiently reproducible ('precise') and accurate. Accuracy can be estimated via the analysis of material that has a known composition (i.e. a 'standard'). 'Analytical precision' should be checked by multiple analyses of single samples. 'Observational precision' can be checked by re-sampling and re-analysing the section of interest (e.g. Fig. 2.1).

Currently four types of periodic or quasi-periodic environmental cycles that produce regular stratigraphic cyclicity have been studied in detail:

(a)   orbital variations generating the tides,
(b)   solar variability affecting the weather and climate,
(c)   orbital variations affecting the weather (daily and annual cycles) and climate (Milankovitch cycles), and
(d)   climatic variations arising from internal processes in the climate system.

Note that climate is typically defined as the mean and variability of the weather as observed during 30-year intervals (Barry and Chorley, 1998). Each of these types of environmental cycle and some of the time series characteristics are discussed in detail in Chapter 6. Missing from this list are cycles related to relatively poorly understood phenomena that might be responsible for some cyclicity observed in stratigraphic

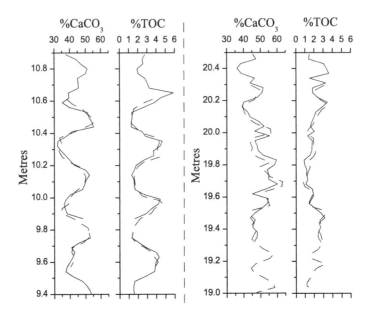

Fig. 2.1 It is not possible to sample the entire Belemnite Marls at a single point on the cliff below Stonebarrow Hill in Dorset. Instead samples were collected from overlapping stratigraphic intervals using exposures about 100 m apart along the gently dipping bedding. The figure shows overlapping series obtained at the 10 m level (left) and at the 20 m level (right). At about the 10 m level a short interval of samples (dashed line) was needed to fill the gap at 9.8 m between long sample runs (continuous lines). The sampling and analyses of the overlapping records revealed good reproducibility of $CaCO_3$ and total organic carbon (TOC) determinations. The offsets between the continuous and dashed lines are partly due to a combination of errors in stratigraphic position during sample collection and errors during measurement of the composition.

records. Principal among these is periodic fault activity (Cisne, 1986; Scholz, 1998; Morley *et al.*, 2000).

2.2   Categories of cyclostratigraphic time series

In mathematics there is a distinction between time series representing 'continuous time' and those representing 'discrete time'. Continuous time refers to mathematical functions that exist at every conceivable point along a time series. Discrete time refers to sequences of numbers existing only at distinct time points. In electronic signal processing continuous time relates to analogue signals and discrete time to digital signals (Ifeachor and Jervis, 1993). Observational time series can be constructed from variables that may be continuous or discontinuous. Continuous variables include most measurements (e.g. $\%CaCO_3$) whereas discontinuous variables include count data (e.g. the number of species of microfossil per sample) or rank (e.g. limestone or shale).

However, regardless of the type of variable used in constructing observational time series, in cyclostratigraphy the observed record can be thought of as derived from what are termed here 'continuous signals' or 'discrete signals'. Consider a sequence of alternating silt and clay laminae. Measurements of average grain size can be obtained using a sample spacing defined by the investigator. A time series of average grain size would be called here a 'continuous-signal record'. However, measurements of successive silt/clay couplet thicknesses would constitute a 'discrete-signal record' because the measurement spacing is dictated by the environmental process that generated the couplets. For example, the couplets of silt and clay laminae might be believed to represent varves and thus each couplet would record deposition during one year. The measurement of couplet thickness can only be obtained at discrete, yearly, intervals, because the 'signal' of couplet thickness is discrete (produced at annual intervals) and not continuous (not produced throughout the year).

2.2.1   Continuous-signal records

'Continuous-signal records' refers here to cyclostratigraphic time series obtained using a depth/thickness scale or a time scale with the sample interval controlled by the investigator. Continuous-signal records include the majority of time series created using sampling intervals that correspond, in time or depth, to more than a year. There appear to be two ways in which continuous-signal records are encoded stratigraphically: (a) indirectly via a carrier wave and (b) directly.

In electronics a radio or electrical carrier wave with a single fixed frequency can be used to transmit information by using a signal that causes a variation in the carrier wave amplitude (i.e. imposed amplitude modulation see Section 1.3, Olsen, 1977). In that case the information is recovered by measuring the carrier wave amplitude (Fig. 2.2). According to Anderson (1986, 1996), for cyclic sediments related to orbital-climatic (Milankovitch) cycles for example (Section 6.9), in some cases it is an annual carrier wave that encodes the annual climatic signal. Modulation of the annual cycle amplitude is imposed by longer period parameters such as variations in orbital tilt (i.e. obliquity).

Normally there is no evidence for varve-like laminations in sediments due to the smoothing of the record of varying sediment composition by burrowing (i.e. bioturbation) and/or laboratory sample homogenization (or a lack of an annual environmental signal to generate varves). As shown in Section 1.3 (Fig. 1.5), simple amplitude-modulated records do not contain oscillations with the beat periods, so smoothing of such signals by bioturbation can remove all evidence for the amplitude modulation (Section 5.3.2). Thus amplitude modulation on its own is not enough to explain the frequent observation of long-period cycles in sediment composition (e.g. Milankovitch cycles). However, some form of partial signal '**rectification**' introduces oscillations related to the beat periods by causing long-period variations in the short-term signal average (Fig. 2.2). Rectification involves a non-linear relationship between the input and output signal amplitudes (Section 5.3.1).

There are many conceivable processes that could cause rectification in the case of continuous-signal records. One of the simplest, suggested by Anderson (1986),

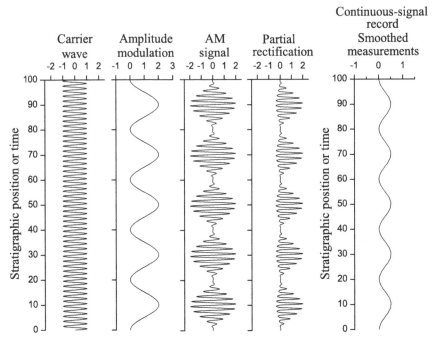

Fig. 2.2 In electronics AM signals are transmitted using carrier waves that have been subjected to imposed amplitude modulation (Section 1.3, Olsen, 1977). In cyclostratigraphy, the carrier wave is often the annual cycle. The AM beat signal can only be preserved stratigraphically if there is at least partial rectification of the carrier wave. For example, consider smoothing the AM signal shown in the middle panel. Ultimately increased smoothing would result in a straight line at the zero level. Hence smoothing of sediment composition, due to processes such as bioturbation, destroys the potential evidence for the annual cycle carrier wave.

However, if there are processes causing rectification (Fig. 2.3) then there can remain a record of the amplitude modulation (beat) frequency even after smoothing has destroyed the evidence for the annual cycle. Continuous-signal records in cyclostratigraphy consist of measurements of some parameter as a function of stratigraphic position or time. The sample interval is largely under the control of the investigator – though constrained by considerations of aliasing (Section 2.4.2).

involves different responses of different parts of the environment. In particular, rectification occurs if, in a two-component system, the different components respond simultaneously or 'in-phase', but with different amplitudes. This is illustrated with a deliberately simplistic model in Fig. 2.3. Two sediment components A and B, having the same densities, could have the same pattern of variation in flux through time (cf. Herbert, 1994). In the example, component A has a smaller amplitude of variation than component B (Fig. 2.3). The total mass of accumulated sediment, composed of A and B only, has the same pattern of variation as each individual component. However, the percentage of A exhibits a partially rectified annual signal. Smoothing of this rectified signal during bioturbation could destroy all evidence for the annual cycle or carrier wave, but leave a small-amplitude record of the long-period amplitude modulation or beat frequencies.

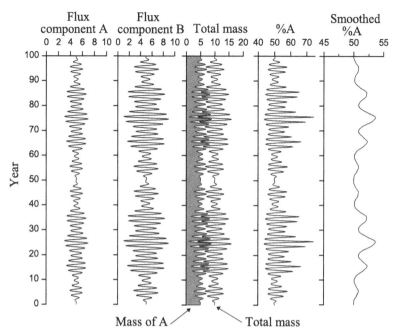

Fig. 2.3 A simple example of the way in which rectification of an amplitude-modulated annual cycle can occur during sedimentation. In this example, the fluxes of two sedimentary components both of unit density record the same amplitude-modulated annual signal, but with different responses (coefficients of variation, Anderson, 1986). The total mass of the two components reveals the same amplitude modulation signal as the components. However, the proportion of component A shows a partially rectified response. Bioturbation results in removal of the evidence from the record of the annual signal, leaving weak longer-term cycles in %A related to the 10- and 50-year beat cycles. Note the change in the scale of the plot of Smoothed %A compared to the other plots.

This simple example illustrates the ease with which environmental and/or sedimentation processes can cause the rectification of continuous-signal records. Anderson (1986, 1996) provides examples of this type of rectification and the generation of long-period cyclicity in varves and previously varved sediments. Anderson (1986) and Herbert (1994) also discuss other types of rectification processes such as out-of-phase relationships between components. Ricken (1993) discusses three-component systems.

Direct encoding of environmental cycles is probably also common. For example, consider the case of Pleistocene sedimentary cycles in deep-sea sediments recording Milankovitch cycles (Section 6.9). There must have been major changes in mean sediment composition (e.g. %$CaCO_3$) over tens of thousands of years that were directly related to changes in the environment – particularly the climate and relative sea level. To this direct signal, a rectified and smoothed annual signal may also have contributed to the changes in mean sediment composition. However, for pre-Neogene sediments it will probably remain difficult to be sure to what extent continuous-signal records result from direct or indirect encoding of the environmental changes.

2.2.2   Discrete-signal records

'Discrete-signal records' refers to cyclostratigraphic time series obtained when the thickness of successive layers (laminae, beds, cycles or growth bands) forms the measured variable, and the layer or cycle number is used in place of the time or depth/thickness scale. For example, in settings where tides or the annual cycle is recorded stratigraphically, a periodic signal dominates the environmental variability. In such cases the lamina or cycle number can be treated as a time scale. On the other hand, if the lamina-generating process is quasi-periodic or aperiodic then of course this is not the case.

Superficially generating time series by using laminae/cycle thicknesses is reminiscent of the situation in electronics where a carrier wave of fixed amplitude is modified by the signal to be sent such that the carrier wave wavelength/frequency is varied (Olsen, 1977). This is called **frequency modulation (FM)**. Spectral analysis of the modulated signal (e.g. of the left-hand panel in Fig. 2.4) would reveal a spectral peak related to the original carrier wave frequency and associated small combination tone peaks that relate to the frequency causing the modulation (e.g. Rial, 1999). The frequencies of the combination tone peaks are the same as for imposed AM (Section 1.3, Fig. 1.5).

However, there are two important differences between discrete-signal records and FM signals. Firstly, the measurements are only obtained at the boundaries of successive oscillations in layer composition rather than throughout each oscillation (compare grain size measurements throughout silt/clay couplets with the measurement of couplet

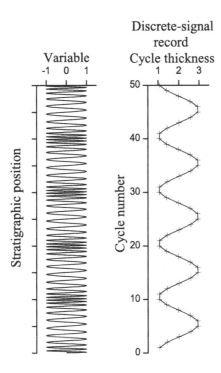

Fig. 2.4 In cyclostratigraphy, discrete-signal records are obtained by measuring the thickness of successive stratigraphic increments (cycles, laminae, tree rings, coral bands, ice layers, cave-calcite layers, etc.). The sample interval of discrete-signal records is dictated by the wavelength of the stratigraphic increments, not by the investigator.

thicknesses). As a result discrete-signal records do not have combination tone peaks related to varying sediment increment thickness (Fig. 2.4). Secondly, the processes generating the variations in sediment composition might be periodic or aperiodic. In either situation the variation in sediment increment thickness usually does *not* provide information about the period of the cycle (with an exception outlined below). For example, the variation in thicknesses of varves does not imply that the length of a year has varied.

Importantly the sample interval used for discrete-signal records is not chosen by the observer, instead it is dictated by the wavelength of the environmental signal. There are three categories of discrete-signal records: periodic, quasi-periodic and aperiodic.

2.2.2a   Periodic discrete-signal records

When the signal controlling the layer formation can be considered to be periodic, each layer represents equal time periods and the time step can be substituted for the layer number. The variations in successive layer thicknesses indicate changes in variables such as sediment flux in sediments, or growth rate in corals, etc. Such variables can be related indirectly to the environment. Conceivably, periodic discrete-thickness records can be obtained from semi-diurnal and diurnal tidally influenced sediments (Section 6.3, Visser, 1980), and sediments recording the annual cycle (Section 6.4) as varves (Boygle, 1993; Anderson, 1996), tidal growth bands in bivalves (Evans, 1972), tidal and annual growth bands in corals (Wells, 1963; Shen *et al.*, 1992), annual tree rings (Stockton *et al.*, 1985), as well as annual light/dark banding in ice sheets (Alley *et al.*, 1997), stalagmite laminae (Qin *et al.*, 1999) and tufa laminae (Matsuoka *et al.*, 2001).

2.2.2b   Quasi-periodic discrete-signal records

When the net accumulation rate is virtually constant, the thickness of successive layers or sedimentary cycles will indicate the period of the cycle-generating mechanism. In such cases, the discrete-signal record allows FM analysis (Hinnov and Park, 1998). For example, de Boer and Wonders (1984), Schwarzacher (1964), Schwarzacher and Fischer (1982), Fischer and Schwarzacher (1984) and Hinnov and Park (1998, 1999) have all argued that the spectrum of variations in the thickness of certain pelagic sedimentary cycles matches that of the variation in the period of the orbital-precession cycle (Section 6.9.1a).

2.2.2c   Aperiodic discrete-signal records

Aperiodic discrete-signal records involving variable time steps between layers include measurements of bed thicknesses when there is no reason to suppose that each bed represents a fixed time interval. The most common example of this type of record concerns the thicknesses of event beds (i.e. turbidites and tempestites, Einsele and Seilacher, 1991), though the frequency of turbidity currents can be controlled by quasi-periodic processes such as cycles in sea level (e.g. Droxler and Schlager, 1985; Reijmer *et al.*, 1991).

2.3   Requirements for the generation of stratigraphic time series

Not every stratigraphic section can be used for obtaining a time series: three conditions must be met as outlined below.

2.3.1   Condition 1 – Consistent environmental conditions

In order to be meaningful, the measured variable must have been determined by the same environmental factor(s) throughout the time interval represented by the data. Hence sedimentological data should be collected from an interval that is free from major facies changes. The recognition of distinct facies can be difficult in some cases, but time series should be obtained from intervals between obvious facies breaks. Establishing the relationship between a particular environmental factor and the measured variable can be problematic, but sedimentological observations and analogy with the modern environment is usually helpful.

Certain stable isotope ratios are especially useful for time-series analyses because they are independent of facies and have global significance (e.g. $\delta^{18}O$ determined using the calcite of benthic foraminifera, Shackleton and Opdyke, 1973). Similarly, estimates of sea surface temperature, derived via transfer functions from microfossil assemblages, are independent of facies, but possess a more local significance (Ruddiman, 1985). Geomagnetic intensity and reversal records can also be used to generate time series that are useful for global correlation (e.g. Guyodo and Valet, 1999; Kent, 1999).

Condition 1 may appear to rule out the analysis of sequence stratigraphic information and long runs of wireline log data from shelf-sea siliciclastic sequences. However, analysis of such information is generally feasible provided that the observed variable is obtained from between major sequence boundaries (Molinie et al., 1990; Worthington, 1990) or if the same set of facies is repeated between sequence boundaries throughout the section of interest (e.g. Naish and Kamp, 1997; Naish et al., 1998). Conceivably sequence stratigraphic data from many localities, used to create a continuous record of regional or global sea-level fluctuations and assigned a time scale, would be amenable to analysis. However, the problem with such an approach has always been obtaining meaningful estimates of the amplitude of sea-level change, free of local tectonic factors, in tandem with a sufficiently reliable time scale.

2.3.2   Condition 2 – Unambiguous variable

The variable recorded must have had an unambiguous relationship with some changing aspect of the environment. An example of a problematic measurement might be total gamma ray counts from a wireline log when the varying contributions from potassium-, thorium- and uranium-bearing phases are unknown. Clearly it would be pointless to use a measurement that was determined by diagenetic processes that were entirely independent of the environmental conditions. In the case of carbonate percentage measurements from a pelagic or hemipelagic limestone-shale sequence, it would be necessary to prove a primary distinction between rock types (e.g. Hallam, 1964; Ricken, 1986; Ricken and Eder, 1991; Herbert, 1993; Munnecke *et al.*, 2001). For the Belemnite Marls, used to illustrate the analysis of real data in the book, there are many lines of evidence indicating both distinct sediment types at the time of deposition and that carbonate contents have only been minimally altered during diagenesis (Sellwood, 1970; Weedon and Jenkyns, 1990, 1999; Munnecke *et al.*, 2001).

Ideally, the possibility that each variable value only relates to one environmental state should also be established. For example, a record of deep-sea calcium carbonate contents would be difficult to interpret if it was known that low values sometimes resulted from periods of sea floor dissolution, and at other times from high fluxes of siliciclastic minerals. In three-component systems (e.g. carbonate, organic matter and siliciclastic minerals) it is important to be sure that only two components dominate the time-series oscillations. When three components form a significant fraction of the sediments, the proportion of one is influenced by variations in the other two variables. However, the series can be modified, for example by using total organic carbon re-expressed on a 'carbonate-free' basis (Ricken, 1993).

Similarly for discrete-signal records the thickness of sediment increments should be related to a single environmental variable. Thus, for example where annual laminae are composed of siliciclastic and biogenic layers, these should be measured and used independently (Bull *et al.*, 2000). Again sedimentological studies and modern analogies are essential for establishing the link between the measured variable and the environment. Unfortunately in some environments, such as carbonate platforms, the lithofacies can exhibit a complex relationship with the environmental factors (e.g. Wilkinson *et al.*, 1997). Nevertheless, in many cases the characteristics of the time series provide strong evidence of a link between the variable and a particular environmental control (Chapter 6).

Digitized qualitative states or rank series such as lithological codes plotted against stratigraphic position represent discontinuous variables (Fig. 2.5, e.g. Weedon and Read, 1995). Discontinuous variables are sometimes examined using Walsh, rather than Fourier, analysis (Section 3.4.6, e.g. Weedon, 1989).

2.3.3 Condition 3 – Thickness–time relationship

2.3.3a Continuous-signal records

For continuous-signal records there must be a relationship between stratigraphic thickness and time (Schwarzacher, 1975). The relationship need not be uniform throughout the data set and the presence of gaps is not insurmountable as will become clear later (Section 5.3.3). For meaningful time-series analysis, increasing stratigraphic thickness must be indicative of increasing time span. In most depositional regimes it is known that thicker sections represent greater time periods, though the longer the time interval, the greater the chance that in detail the section is incomplete (Section 5.3.3, Sadler, 1981; Anders *et al.*, 1987; Ricken, 1991a). Whether a thickness–time span relationship can be demonstrated depends on the circumstances. In deep-sea and lacustrine settings

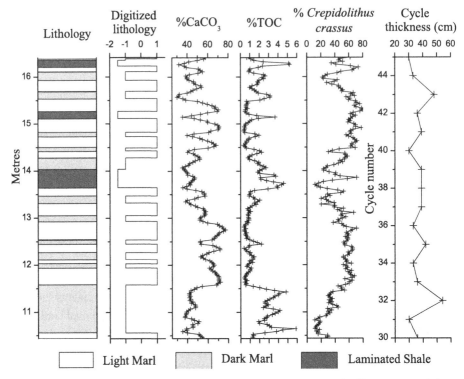

Fig. 2.5 A single stratigraphic section can yield a variety of different types of time series. Here five time series are illustrated for a short interval of the Belemnite Marls (cf. Fig. 1.10). Continuous-signal records: a square wave (digitized lithology), two geochemical measurements (TOC indicates total organic carbon) and the proportion of a nannofossil species. The nannofossil data were kindly provided by Ben Walsworth-Bell. Discrete-signal record: couplet cycle thicknesses (measured between carbonate maxima).

it can normally be assumed when event beds are absent. Yet in shallow siliciclastic settings accommodation space and facies influence sediment thickness. In platform carbonates it is not even clear whether there is always a consistent relationship between accommodation space and sediment thickness (Boss and Rasmussen, 1995; Wilkinson *et al.*, 1999).

For short-period cyclicity, where tidally or annually generated laminae or growth bands are absent, radiometric dating is required before the data can be examined against a time scale. Such time scales require careful evaluation (e.g. Andrews *et al.*, 1999).

Many Late Cenozoic deep-sea sediment records can be dated by matching benthic $\delta^{18}O$ values to a global $\delta^{18}O$ standard. This standard has been 'orbitally tuned' to calculated orbital variations extending back to 6 million years before present (6 Ma BP, Section 6.9.3). For older Late Cenozoic sections, tuning based on carbonate, magnetic susceptibility and reflectance records has allowed absolute dating of biostratigraphic and magnetostratigraphic events, and calibration of the marine strontium-isotope curve back to about 30 Ma BP (Section 6.9.3). The 'astronomically calibrated' records are useful for studying environmental cycles with periods of tens of thousands of years to a few million years.

Most pre-Oligocene stratigraphic sections currently lack a time scale of sufficient detail to identify gaps and to characterize the accumulation rate variations over periods of thousands to hundreds of thousands of years. As a result, the thickness–time span relationship is important when testing continuous-signal records with a depth or thickness scale for regular cyclicity. In this connection Schwarzacher (1975) provided an important proof of **Sander's Rule** (Sander, 1936). This indicates that an environmental cycle that occurs regularly in time might be expressed stratigraphically (i.e. in thickness) as either regular or irregular oscillations. But it is exceedingly unlikely that irregular environmental oscillations will be expressed in the rock record as regular sedimentary cyclicity. Consequently, the detection of sedimentary cycles of constant thickness implies a regular environmental cycle. Lack of regularity in thickness cannot necessarily be interpreted as implying a lack of regularity in environmental changes. It could mean that no regular oscillations occurred in the environment, or that interruptions and distortions of a regular environmental cycle prevented the formation of a regular stratigraphic signal (Sections 5.2 and 5.3).

For environmental changes lasting millions to hundreds of millions of years, the data can be related to the geochronometric scale. For example, data concerned with rates of extinction were analysed using dates from successive stages (Sepkoski and Raup, 1986; Raup and Sepkoski, 1988; Sepkoski, 1989). However, the small number of data points combined with problems with the time scale used apparently led to biased results (Stigler and Wagner, 1987, 1988). Analysis of extinction and rates of evolution with time-series methods remains topical (Prokoph *et al.*, 2000; Kirchner, 2002; Peters and Foote, 2002).

2.3.3b   Discrete-signal records

For periodic discrete-signal records the successive sediment or biogenic growth increments must usually represent equal, and preferably known, time intervals. However, it can be extremely difficult to establish that this is the case. In settings recording varve-like laminae, in the absence of sediment-trap data, it must be borne in mind that laminae are not necessarily produced annually (e.g. Halfman and Johnson, 1988; Crucius and Anderson, 1992; Christensen *et al.*, 1994; Hagadorn, 1996). Thus independent dating methods, such as short-lived radio-isotopic studies, are required for establishing an annual period for lamination before a discrete-signal record can be constructed. Sub-annual tree rings create spurious oscillations that can be corrected by cross-matching multiple records (Stockton *et al.*, 1985; Schweingruber *et al.*, 1990). Likewise in mixed semi-diurnal and diurnal tidal settings (Section 6.3) different laminae represent either half a day or a day (Archer and Johnson, 1997). On the other hand, sometimes the characteristics of periodic discrete-signal records are diagnostic of a particular type and period of environmental signal (e.g. Archer, 1996).

In sediments recording Milankovitch cycles, meaningful quasi-periodic discrete-signal records can only be constructed when just a single orbital cycle is involved and mean accumulation rates are nearly constant (Herbert, 1994; Hinnov and Park, 1998, 1999). For such strata, variations in cycle thickness can be treated either as a function of cycle number (the 'metronomic method' of Herbert, 1994) or as a function of stratigraphic position (the 'stratigraphic method'). Using cycle number, the quasi-periodic discrete-signal records reveal regular long-period variations in the primary cycle period (e.g. Figs. 12 and 19 of Herbert, 1994). However, using the stratigraphic method means that the variations in cycle thickness are incorporated into the scale that defines the position of the cycle thickness measurements. Consequently, the combination tones caused by the frequency modulation appear in the spectra. Not surprisingly this means that the spectra are much harder to interpret (Herbert, 1994).

In discrete-signal records the relationship between thickness and time is more complex than for continuous-signal records. For successive stratigraphic increments (sediment laminae, growth bands, etc.), the variations in thickness represent changes in the environmental variable of interest rather than time. However, increment thicknesses are partly related to time when multiple stratigraphic increments are considered.

The presence of event beds, within a sequence used to generate a continuous-signal record, complicates the generation of time series. This is because the thicknesses of individual turbidites and tempestites are more likely to relate to storm intensity, earthquake magnitude, or the time interval since the last event, rather than to the time taken for deposition. If such beds are present, they may need to be artificially 'subtracted' from the section prior to time-series analysis, but this presupposes that they can be readily identified.

Event beds, treated separately from the background sediment, can of course be used to create their own aperiodic discrete-signal records. If the event beds interrupt continuous-signal records, the thickness of the background sediment can be used to indicate relative time span between events. For discrete-signal records interrupted by event beds, recurrence time is indicated by the number of discrete sediment increments. The thickness of the event beds can then be used to indicate event magnitude. Event beds used to form aperiodic discrete-signal records are sometimes extremely informative (e.g. Foucault et al., 1987; Haak and Schlager, 1989; Weltje and de Boer, 1993; Beattie and Dade, 1996).

2.4   Sampling

Mathematicians are used to dealing with functions or oscillations in continuous time which can, in theory, be sampled arbitrarily closely. Fourier analysis, the description of time series in terms of the component sine and cosine waves, was developed around functions through the use of integrals. However, the analysis of observational time series required modification of the mathematical methods so that summation is used in place of integrals. This is because the data are obtained by sampling and therefore exist in discrete time.

2.4.1 Sample intervals and power spectra

Since stratigraphic records are obtained from sediments or fossils using discrete sample values, rather than from theoretically continuous mathematical functions, there must be differences in the information that these two types of time series can convey. It would only be possible for observational time series to contain information about truly instantaneous changes in the measured variable if values were collected using infinitesimal sample intervals. Instead, there is a limit to the highest frequency information that can be obtained from real data. This limit is called the **Nyquist frequency** (or **Nyquist critical frequency**) which is defined as $1/(2 \times$ sample interval). Some texts use the term **folding frequency** in place of Nyquist frequency. This definition assumes that the samples are obtained at some constant time or thickness/depth interval. The Nyquist limit arises because the smallest sine or cosine wave which can be represented or reconstructed using discrete samples has two values per oscillation (i.e. the limiting period/wavelength is twice the sample interval, Fig. 2.6). For example, a sample interval of 3000 year (3 ka) yields a Nyquist frequency of 0.167 cycles per thousand years. Obviously the larger the sample interval the lower the Nyquist frequency. In order to compare power spectra, many of the more mathematical texts assume a sampling

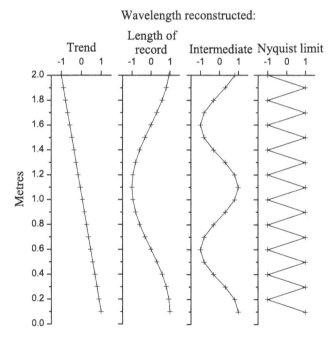

Fig. 2.6 Time series based on sampling or discrete observations can record the presence of a wide range of oscillation wavelengths or frequencies. This range of wavelengths is restricted at short wavelengths or high frequencies by the sampling or Nyquist rate. In this case the sample interval of 0.1 m implies that the shortest oscillations which can be recorded have a wavelength of 0.2 m (2 × sampling interval). This wavelength corresponds to the Nyquist frequency (= 1/2 × sampling interval).

interval of one so that all the spectra illustrated have frequency axes running between 0 and 0.5.

2.4.2   Sample intervals and aliasing

Observational time series are obtained in one of two ways. Firstly, a value might be determined over a short time or stratigraphic interval, then no information is gathered for some interval, another value is obtained and so on. This corresponds to **spot sampling** (also called **point sampling** and **burst sampling**) and is a method of data collection that leaves a certain proportion of the signal unsampled. Spot sampling is often used to generate continuous-signal records. Alternatively, values are obtained as averages spanning the whole length of the sample interval. In other words, the quantities have been established as non-overlapping averages that cover the whole of the time interval represented by the data. This method is called **channel sampling** (or **continuous sampling**), and corresponds to the collection of some types of geophysical data used in wireline/downhole logs, in other words some continuous-signal records and the majority of discrete-signal records.

The implications of using spot or channel sampling are rather different. With spot sampling, the critical issue is to ensure that the samples record information about the highest frequency parts of the record being examined. When high-frequency components are incorrectly recorded the result is called **aliasing** and the time series is described as **aliased** (Priestley, 1981). Aliasing occurs when there are features in the cyclostratigraphic signal that have frequencies higher than the Nyquist frequency. This simply means that any small-scale oscillations are not sampled at least twice so giving rise to spurious low frequencies in the final time series (Fig. 2.7). Aliasing is encountered for example when watching television and film where wheels appear to move too slowly, or even backwards on vehicles. Pisias and Mix (1988) provide examples of aliasing in deep-sea sediments. Wunsch (2000a) argued that a spectral peak indicating regular millennial-scale cycles (Section 6.8), in oxygen isotope records from the Greenland ice sheet, actually results from aliasing of the annual cycle. However, this has since been contested (Alley *et al.*, 2001; Meeker *et al.*, 2001).

In theory, aliasing can be avoided by using a sample interval that is less than half the period/wavelength of the smallest oscillation in the signal being sampled. Channel sampling incorporates an averaging or smoothing of the signal which suppresses the highest frequencies in the signal so that aliasing is usually avoided. In mathematical terminology channel sampling incorporates an '**anti-aliasing filter**'. However, the problem with channel sampling is that, although aliasing is avoided, potentially interesting small-scale features of the signal may be smoothed out. Discrete-signal records are normally effectively composed of channel sample observations. However, discrete laminae within varves, formed during one season rather than throughout the year, would allow spot sampling (e.g. Bull *et al.*, 2000).

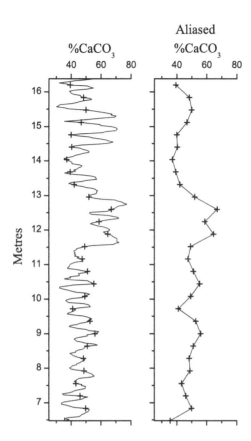

Aliased
%CaCO₃

Fig. 2.7 If the sample interval is too wide for the scale of oscillations being examined, the result is an aliased signal. Aliased time series contain oscillations that are spuriously long. In the illustration part of the Belemnite Marls carbonate record (left) has been reconstructed using samples at 36 cm (indicated using plus symbols). The resulting aliased time series (right) contains long-wavelength oscillations that are partly an artefact of the inappropriate sampling.

Usually the nature of the variability of a cyclostratigraphic signal is unknown prior to sampling. However, if bulk sediment composition is to be determined, it is often the case that fluctuations in colour will help reveal the smallest scale of variation. In other cases, such as geochemical (e.g. stable isotopic) or micro-palaeontological sampling, it is advisable to perform a pilot study using very closely spaced samples from a representative stratigraphic interval. Unfortunately, it is difficult to be sure that the pilot study interval really is representative. Ideally the time series should exhibit smoothly varying successive values. If the sample interval is slightly too large (i.e. there is some aliasing), the resulting time series is very erratic or 'spiky', possessing many sharp peaks and troughs defined by single points rather than smoothly varying values (Fig. 2.8).

Unfortunately the presence of many oscillations defined by just two points could mean that the signal has not been aliased or that aliasing has occurred. As a result a minimum of four points per major oscillation is desirable despite the definition of the Nyquist limit as two samples per cycle. Fairly frequently the oscillations of interest are not simple sinusoids. They may instead consist of broad troughs and narrow peaks or a saw-toothed shape or square waves (Section 5.2.4). In these cases the only way to correctly characterize the shape of the waveforms is by taking many more than two

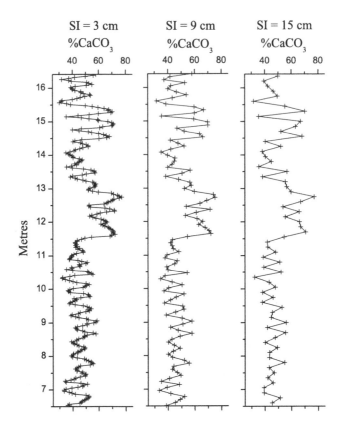

Fig. 2.8 Part of the Belemnite Marls carbonate record with three different sample intervals (SI). The record reconstructed using 9 cm sampling contains all the expected oscillations. On the other hand, the time series based on 15 cm sampling has missed several of the oscillations (at the 11.8, 13.0 and 16.0 m levels). Note that the 15 cm time series is very 'spiky' with just two values defining each oscillation in most cases. This 'spiky' character appears to indicate that the sampling is at the Nyquist limit of the signal (Fig. 2.6) whereas the signal is actually already aliased. Therefore, it is best to have at least four values defining the shortest oscillation (e.g. middle time series illustrated here). In general it is preferable to have oscillations defined by values that change smoothly from point to point between peaks and troughs. Indeed Herbert (1994) recommended at least eight samples per oscillation to ensure that the shape of the signal is well characterized.

samples per oscillation. Herbert (1994) suggested a minimum of eight samples per smallest oscillation. An example of suspected aliasing is provided by the topmost data from the Belemnite Marls (Fig. 2.9).

In the past some sedimentologists have chosen to obtain samples solely at the centre of every bed that is' visible in the field. This is bad practice for a variety of reasons. For example, some subtle lithological variations may not be detectable by eye in the field so important oscillations could be missed. As just explained, the rather 'spiky' nature of the time series, which results from this method of sampling, cannot be assumed to be free of aliasing (Fig. 2.8). Sampling the centres of beds also produces

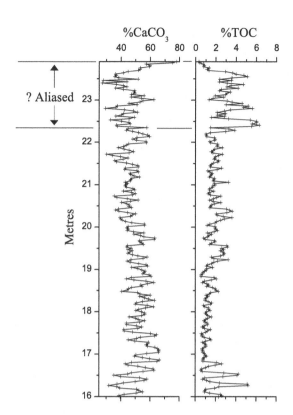

%CaCO₃

%TOC

Fig. 2.9 For most of the Belemnite Marls time series, each oscillation is defined by many points. However, above the 22.3 m level both records take on a rather 'spiky' character suggesting that aliasing has occurred (cf. Fig. 2.8).

irregularly spaced data. This will usually necessitate interpolation (Section 2.4.3) to produce regularly spaced values, but interpolation of very 'spiky' data is unlikely to be successful. In palaeoceanographic work, data are often collected uniformly in depth along a deep-sea sediment core and usually necessitate interpolation when placed on a time scale. Interpolation is usually much more reliable if the irregularly spaced data vary smoothly (i.e. successive values increase and decrease in short runs rather than oscillate from high to low values at successive points). Hence it is much better to chose a sample interval that is considerably smaller than the thinnest visible bed or layer and then obtain uniformly spaced data, rather than to restrict the sampling to visually obvious maxima and minima in sediment composition.

Data that are suspected of being aliased should not be used for time-series analysis. The reason is that incorrect interpretations will follow from incorrectly sampled data – in particular there will be contamination of the low-frequency part of the signal by the aliased high-frequency components. In practice, the degree to which this contamination is a problem is dictated by the real spectrum of the signal. If the high-frequency components are of very small average amplitude (power) compared to the lower frequencies, their aliasing will present little difficulty. If the high frequencies have power that is a substantial fraction of the low frequencies, the aliasing will be problematic.

This is illustrated using the Belemnite Marls data where successively larger sample intervals have been used to generate power spectra (Figs. 2.10 and 2.11). The larger

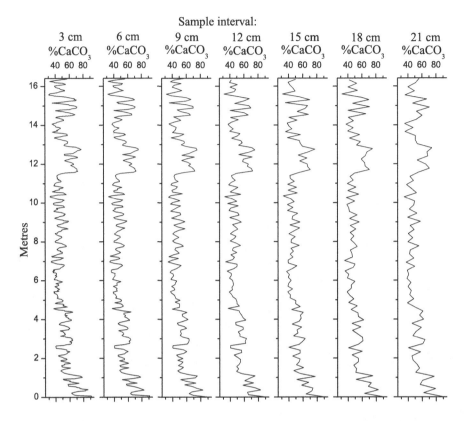

Fig. 2.10 The calcium carbonate composition of the lower two-thirds of the Belemnite
Marls reconstructed using various sample intervals.

sample intervals were obtained by regularly skipping over a fixed number of samples
in the original data – a process called **decimation**. Decimation normally results in
aliasing. However, in this case it is not until very large sample intervals are used
(15 cm or larger) that the low-frequency part of the spectrum is severely distorted
by aliasing. Unfortunately, one cannot know whether the data have been aliased in a
particular time series obtained by spot sampling, since the spectrum beyond the Nyquist
frequency is unknown. For this reason, it is best to conduct a pilot study first using
a very small sample interval from a selected section, and then inspect the resulting
time series to guide the choice of sample spacing for the main study. If in doubt,
it is usually best to collect information using the smallest sampling interval that is
practicable.

Fortunately, spot sampling in sediments is often feasible because of the effects
of bioturbation. Mixing by burrowing organisms tends to act as an anti-aliasing filter, because centimetre- to millimetre-scale fluctuations are smoothed out (Dalfes
et al., 1984). The result is that in many sedimentary successions, spot samples can
usually be collected at quite large separations without causing aliasing. Of course, in

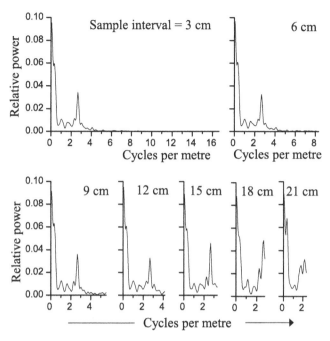

Fig. 2.11 Power spectra corresponding to the time series illustrated in Fig 2.10. Note that at longer sample intervals the Nyquist frequency decreases. Theoretically all sample intervals cause some aliasing, but it is only sample intervals that alias large-amplitude components of the signal that have a major effect on the spectrum. In this case the low-frequency parts of the spectra are essentially identical for sample intervals of 3, 6, 9 and 12 cm. At sample intervals of 15 cm and longer, aliasing introduces significant spurious long wavelength oscillations into the time series. This has the effect of distorting the low-frequency part of the spectrum.

laminated sediments, bioturbation is absent or minimal. Consequently, if continuous-signal records are to be obtained using sample intervals that are independent of the laminae, some method of averaging or smoothing the data is needed. This can be achieved by homogenization of rock samples collected over complete sample intervals (channel sampling) or by smoothing very high-resolution optical data (Ripepe *et al.*, 1991; Schaaf and Thurow, 1997).

2.4.3 Missing values and irregular sample intervals

It is usually desirable to obtain data using a constant time interval or spacing since most computer algorithms are designed assuming that this is the case. Where data have not been collected with fixed sample intervals, the treatment of the time series depends on the type of spacing.

If most values are collected at equal time intervals or are equally spaced, but a few values are entirely missing, interpolation can be used. Interpolation is used to provide

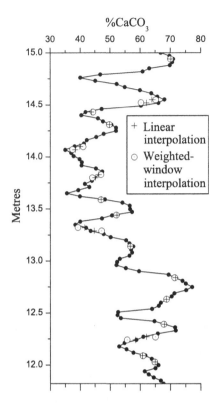

Fig. 2.12 Interpolation can be used to 'reconstruct' missing values or to create values at constant spacing from irregularly spaced samples. Two types of interpolation are illustrated here for part of the Belemnite Marls data. A random number table was used to select and remove about 10% of the values that were then treated as missing. These were reconstructed using linear interpolation (plus symbols) and a Gaussian weighted-window interpolation (circles with dots). The original measurements are indicated by filled circles.

Linear interpolation merely uses a straight line between remaining observations to calculate the interpolation values. This is a simple but crude procedure because time series tend to possess curved rather than straight lines. It can introduce spurious high frequencies into the data and works poorly where there is a lot of high-frequency noise. In the weighted-window procedure, a short section of data or window, centred on the position of the missing value, is used to calculate the interpolation value by assigning greater importance (numerical weight) to observations nearer to that position. This produces smooth curvilinear interpolations, which is particularly useful when data contain a lot of high-frequency noise.

In this example, around half the missing values are single values that were successfully reconstructed by both methods. The rest of the missing values occur in pairs and, not surprisingly, the interpolations were less successful in these cases (e.g. near the 14.5 m level). However, comparing the position of the interpolations with the removed values, the linear interpolation was slightly more successful than the weighted-window procedure. The success of the weighted-window interpolation might have been improved by using a shorter smoothing window.

estimates of the missing values based on adjacent data. The success of interpolation depends strongly on the shape of the time-series oscillations and the algorithm employed
(Yiou *et al.*, 1996). Press *et al.* (1992) provide a variety of routines for interpolation. If
values change rather smoothly from point to point, a simple linear interpolation may
be satisfactory though this method can introduce spurious high-frequency information (Fig. 2.12). More erratic variations may require a weighted-window approach.
In such cases, values closer to the missing value are considered as more useful for
the interpolation (given greater 'weight') than values further away. Unless deliberate
smoothing of the data is required, only a small length of data, which defines the length
of the smoothing window, is used for each calculation. Smooth interpolation can also
be achieved via cubic splines (Yiou *et al.*, 1996).

Sometimes time series can only be obtained with variable sample intervals, for
example when the variable of interest is bed thickness relative to stratigraphic position (rather than relative to bed number). Two approaches are available in these
cases. Firstly, interpolation can be used to generate an entirely new set of uniformly
spaced values. However, the choice of the new sample interval is important. If the
interpolation interval is too wide, the original data are effectively over-smoothed and
the spectrum lacks the expected power at higher frequencies. Using a very small interpolation interval does not add high frequency information. Instead, the resulting
time series would have a large number of smoothly varying values, and the spectrum would have an unrealistically high Nyquist frequency. A common and sensible
choice is to use an interpolation interval close to the mean spacing of the original
data.

Interpolation inevitably causes smoothing of the data, thereby suppressing high-
frequency peaks (Section 5.4.2, Schulz and Stattegger, 1997). Thus an alternative
approach to irregularly spaced data is to use a time-series analytical technique specifically designed for such series. Algorithms exist for spectral and cross-spectral analysis
of unevenly sampled data based on the so-called Lomb–Scargle method (Press *et al.*,
1992; Schulz and Stattegger, 1997). Although undoubtedly useful, it should be borne
in mind that several other methods of time-series analysis require uniformly spaced
data.

If the gaps between data points are much larger than the scale of the largest oscillations of the signal being sampled, neither interpolation nor special algorithms are much
help. In these cases, the time series should be split into a collection of short records.

2.5   Chapter overview

■ All cyclostratigraphic time series belong to one of two categories. Continuous-
signal records have sample intervals chosen by the investigator. Discrete-signal
records have sample intervals dictated by the periodic, quasi-periodic or aperiodic phenomenon generating the stratification (e.g. the annual growth bands in
corals, trees, etc. and the laminations in varved sediments).

■ Meaningful cyclostratigraphic time series can be generated when:

   (a) the environmental conditions remain reasonably consistent,

   (b) the variable observed/measured provides an unambiguous proxy for the state of the environment, and

   (c) for continuous-signal records there is a relationship between stratigraphic thickness and time span.

■ The design of a sampling programme for generating continuous-signal records must consider the issue of aliasing.

Chapter 3

Spectral estimation

3.1 Introduction

As mentioned in Chapter 1, generating a spectrum is often the first step in the analysis of cyclostratigraphic time series. Unfortunately, researchers sometimes use one of the widely available computer programs without regard to the theory or methodology being implemented. A working understanding of the main issues will reduce the chances of making mistakes and enable the investigator to make the most of the data. A primary use of the spectrum is for the detection of periodic or quasi-periodic components in a time series, which are indicated by peaks that can be distinguished from background noise (i.e. significant spectral peaks). Note that the periodic or quasi-periodic components are referred to here as 'regular cycles' rather than just 'cycles' for reasons explained in Section 1.3.

This chapter is divided into four sections that consider: (a) processing time series prior to spectral analysis; (b) the background to the generation of power spectra; (c) the main methods used to generate spectra; and (d) some of the issues concerning distinguishing spectral peaks from background noise. The practical interpretation of power spectra is also addressed in Section 5.4.2.

It should be kept in mind that it is impossible to use cyclostratigraphic time series to obtain the theoretical or 'true' spectrum. The reasons for this will become clear in this chapter. Different spectral techniques employ different methods to obtain a 'best guess' as to the true spectrum. In fact the correct term for the procedures described here is '**spectral estimation**' and the values that describe the shape of the spectrum are termed **spectral estimates**. There has been a tendency for individuals to describe a particular method of spectral estimation as the 'most satisfactory' or 'superior' to other methods. Do not be fooled by such prejudices. Whilst it is true that the periodogram method, described shortly, is unsatisfactory, all the other common methods have their merits and in general most methods give similar results. The similarity of the different types of spectra can be judged by their use on the same set of data from the lower part of the Belemnite Marls (using the number data points, $N = 512$) in Section 3.4. The aim here is to introduce the most important considerations and then leave the use of specialist methods to the choice of the reader in the context of the particular data set being studied. Occasionally, it has been suggested that the best strategy is to employ a succession of spectral estimation techniques – in other words, different techniques are based on different compromises (Pestiaux and Berger, 1984a; Hinnov and Goldhammer, 1991; Yiou *et al.*, 1996). The Appendix includes a list of sources of algorithms for spectral estimation.

The explanations that follow do not include the formal mathematical treatments that are common to most texts on spectral analysis. For those interested in such approaches Bloomfield (1976) is an excellent starting point and Priestley (1981) provides a thorough mathematical introduction. A highly readable and down-to-earth style, with a digital signal processing bias, is provided by Ifeachor and Jervis (1993). For an up-to-date, well-illustrated, but very detailed mathematical account consult Percival and Walden (1993). The subject is replete with technical synonyms so I have generally tried to stick to the terminology adopted by Priestley (1981), and cite the alternatives as each term is introduced.

3.2 Processing of time series prior to spectral analysis

There are several methods used to pre-process time series, so that the resulting spectra have shapes that are more easily interpreted or less influenced by mathematical artefacts. Some methods (mean subtraction or linear detrending discussed below) are applied to all time series prior to spectral analysis, whereas others are only used for certain types of data. Because the methods described in this section are considered fairly standard, in publications it is normal for pre-processing to be performed without illustration of the results.

3.2.1 Mean subtraction

In general, spectra are designed to indicate the mean square of the deviations from the zero line for different frequency components. This means that, even if no other processing is needed prior to spectral analysis, the mean of the whole time series is subtracted from every value. This leaves a series oscillating positively and negatively around the zero line. Such data are sometimes referred to as **centred**.

3.2.2 Ergodicity, stationarity and detrending

All physical systems exhibit a degree of entirely unpredictable, random behaviour. This means that two time series used as 'snap shots' of the behaviour of the system will never be exactly the same. Mathematically each time series used to describe precisely the same system is called a **realization**. If a single series is long enough to be used as representative of all possible realizations, it is described as **ergodic** (Priestley, 1981). For the majority of cyclostratigraphic time series there is no choice but to hope that the methods that rely upon **ergodicity** will still provide meaningful information even if the data are not ergodic. The issues connected with deciding how long a time series needs to be are tackled later (Section 5.4.1).

In practice one usually only has one realization to examine, because each time series represents a unique record of the local environmental history. However, $\delta^{18}O$ measured in the tests of benthic foraminifera, marine $^{87}Sr/^{86}Sr$, geomagnetic intensity and magnetic reversals exemplify measurements that have been treated as of global significance and thus multiple time series can be compared (e.g. Imbrie *et al.*, 1984; Prell *et al.*, 1986; Guyodo and Valet, 1999; Martin *et al.*, 1999). $\delta^{18}O$ records obtained from geographically closely spaced ice cores in Greenland allowed comparison of multiple records of atmospheric temperature (Johnsen *et al.*, 1997). Tree ring data sets are also based on information collected from multiple trees from a wood or forest – each studied series usually represents the average of multiple detrended realizations from different trees (Stockton *et al.*, 1985; cf. Esper *et al.*, 2002). In these few cases a new time series can be created by **stacking**, or averaging, the individual records. Stacking is used to reduce variability in the data due to measurement and other errors. By treating the variable of interest as signal and the measurement errors as noise, stacking improves the '**signal-to-noise ratio**' by a factor of $\sqrt{n}$, where n represents the number of records (Lowrie, 1997). In other words, doubling the signal-to-noise ratio requires stacking four individual time series.

Spectra illustrate the average distribution of power as a function of frequency over the interval represented by the complete time series. Clearly if the time series changes its characteristics significantly over the length of interval studied, the spectrum is not easy to interpret. Put another way, spectra are constructed mathematically as though the time series is unchanging or **stationary**. Since noise is always present (due to measurement errors and unpredictable variability in the environment), not all aspects of a time series can remain constant so to some extent all cyclostratigraphic time series are **non-stationary**. But in fact, for the spectrum to be interpreted as a useful description of the time series, all that is usually required is that the mean and variance of the data remain essentially constant ('**weak stationarity**', Priestley, 1981, 1988).

If the mean value of a time series shows a **trend**, i.e. a progressive increase or decrease along the length of the data, the series is non-stationary. In signal processing terminology, by analogy with electronics, the trend is sometimes referred to as the **DC** (direct current) **component** and the rest of the detrended oscillatory time series as the **AC** (alternating current) **component**. The trend in the mean needs to be removed prior to spectral analysis by **detrending**. The Fourier transform used for generating spectra, explained below, treats the data, which are of finite length, as though they are repeated infinitely. With a trend present this means that after transformation the data appear to have a cycle with a length equal to the length of the data (Fig. 3.1).

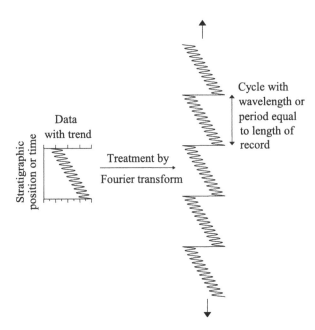

Fig. 3.1 When the mean of a time series shows a progressive increase or decrease, or in other words a trend, the data are described as non-stationary. The Fourier transform, used in generating spectra, treats the data as though they are repeated indefinitely. Thus mathematically data containing a trend appear to have a cyclic component with a wavelength or period equal to the length of the data. Consequently, removal of linear trends via linear regression is routine in spectral analysis.

A common form of detrending is to fit a linear least-squares regression line to the data and then to subtract the line, point by point, from the series. This prevents distortion of the low-frequency part of the spectrum (Figs. 3.2 and 3.3). The detrending eliminates the 'distortion' of the spectrum due to the DC component. A trend in mean values is often explicable in terms of some long-term changes in the environment being monitored. If the trend is not linear, but can be approximated using a first-order polynomial, one can treat the trend as part of a cyclic component with a wavelength exceeding the length of the data (called a **hypercycle** by Dimitrov *et al.*, 1998). Normally, since hypercycles show monotonic increases or decreases rather than oscillations, time-series analysis cannot be used to make useful statements about this component of the record (cf. Mann and Bradley, 1999). For instance, would a trend reverse just beyond the ends of the series or go on much further? Detrending eliminates the power at zero frequency so this value is often left off the spectral plot. Subtracting the trend means that the data oscillate around the zero line so that effectively mean subtraction is automatically performed by detrending.

When the data have a non-stationary mean due to a step increase or decrease in average values, simple linear detrending does not help. Instead the unwanted low-frequency component in the spectrum can be removed by splitting the data into two segments at the level of the step (preferably with an understanding of the cause of the step).

Non-stationary variance can have manifestations as for example: (a) a long-term increase or decrease in average amplitude, (b) intervals of unusually high or low mean amplitude, or (c) changes in the dominant wavelength or periods of variation (Fig. 3.4). A typical case of the third type of non-stationarity is where, for a time series based on a thickness scale, there is a change in mean accumulation rate despite the persistence

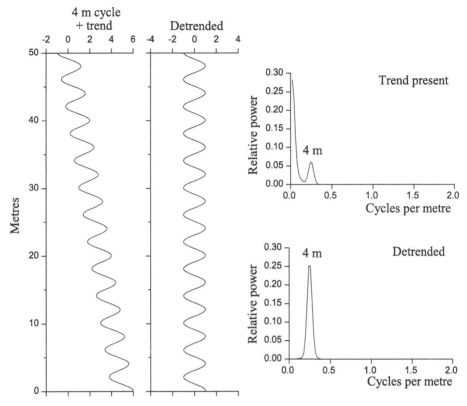

Fig. 3.2 Figure showing how the spectrum is contaminated by power at the lowest frequencies if a record is not detrended prior to analysis.

of a regular environmental cycle in time (Fig. 3.4, Section 5.2.1). Sometimes changes in mean variance are visually obvious in a time series. If there are no visually obvious changes in variance it is advisable to check for stationarity by subdividing the time series and inspecting the corresponding spectra for consistency (Priestley, 1988). It is rare for a time series with non-stationary variance to be manipulated to produce stationarity in the absence of a detailed knowledge of the characteristics of the environmental system (Schwarzacher, 1975). Instead the data are examined using spectra from non-overlapping stationary segments, or by using spectra from overlapping segments when the variance is assumed to change 'gradually' (see Section 4.2). There are techniques for studying non-stationary time series sometimes using specially designed mathematical models, but these are not considered further here (e.g. Priestley, 1988; Gershenfeld et al., 1999; Young, 1999).

The complete sets of Belemnite Marls carbonate and organic carbon data are not stationary with respect to the variance. The decimetre-scale couplets decrease in thickness above the 16.41 m level (Fig. 1.10, Section 4.2). Nevertheless, these data, used for illustrating the various techniques discussed, are stationary with respect to the variance below the 16.41 m level.

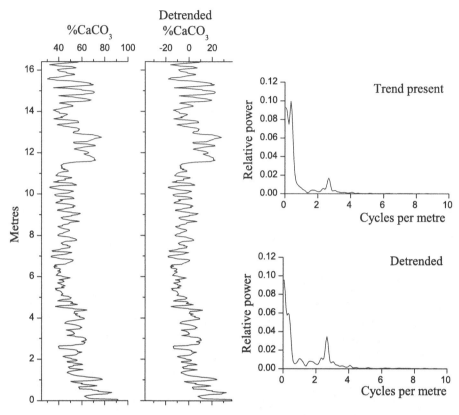

Fig. 3.3 The Belemnite Marls has a comparatively small trend of decreasing carbonate contents up-section (i.e. a decrease that is similar in size to the amplitude of the couplet-scale cycles). Nevertheless, if the data are not detrended prior to spectral analysis, there is a substantial distortion of the spectrum.

3.2.3   Outlier removal and the unit impulse

When a time series has a few points that have values well beyond the range of the other values it is said to contain **outliers**. In order to understand the effect of outliers on the spectrum the unit impulse needs to be introduced.

The **unit impulse** is a special form of 'real' time series (i.e. consisting of discrete-time observations) consisting of zeroes with a single value of one ('unity'). (In mathematical treatments of 'theoretical' time series – i.e. based on continuous time – one uses an impulse with an infinitesimal width called a **delta-** or **Dirac-function**). The spectrum of a unit impulse is a horizontal line (Fig. 3.5). This spectral shape arises because the unit-impulse time series decomposes, during Fourier analysis, into regular oscillations representing every frequency with, in the example illustrated, every component having equal amplitude. The component regular waves must be aligned (i.e. possess particular phases) so that all the peaks and troughs cancel each other, except for the peaks that are aligned with the unit impulse itself (Fig. 3.5). Note that sometimes outliers can produce **ringing** (alternating high and low power values) at higher frequencies (Percival and Walden, 1993).

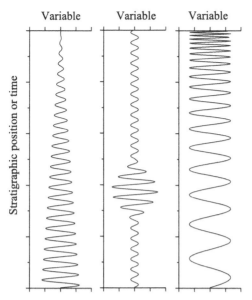

Fig. 3.4 Non-stationary variance in a time series can occur in a variety of ways. The local variance can show a gradual increase or decrease (left), a short interval of large-amplitude oscillations (middle) or a shift from one frequency range to another (right). In all these cases a spectrum, which plots average variance (or squared average amplitude) against frequency, would provide a misleading description of the data. In order to be meaningful, spectra should be calculated where the mean variance is more or less stationary. In observational time series non-stationary variance may be difficult to detect visually. To check whether the variance is stationary the spectra from subsections of the data should be compared.

This result means that if an observational time series has several very large or very small (or negative) values there is a spurious background to the spectrum which is also flat (Fig. 3.6). The spurious background is caused by the outliers effectively adding unit impulses to the rest of the data (Fig. 3.5). The more anomalous the outlier(s), the larger the potential distortion of the expected spectrum. Some time series algorithms are designed to cope with missing values or outliers (denoted using a substitute value of, e.g. −999) which are disregarded during the computation. In general though, outliers should be replaced by either interpolated values or the local mean value. Elimination of outliers presupposes an objective method for their identification (e.g. known analytical problems).

3.2.4   Pre-whitening

For time series with spectral power that drops off towards high frequencies, it is sometimes useful to modify the time series to produce a flatter spectrum (Priestley, 1981; e.g. Hays *et al.*, 1976). Such a sloping spectrum is often said to describe **red noise** since red light is dominated by low frequencies (Section 3.3.1). Red noise spectra have a large **dynamic range**, i.e. the power values are generally large in one part of the spectrum and generally very small in another part. Pre-whitening generates a time series

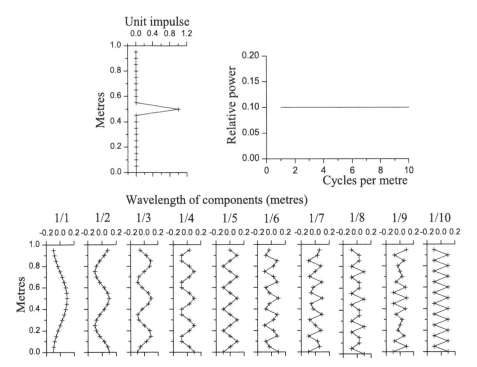

Fig. 3.5 The unit impulse is a time-series composed of zeroes with a single value of one (or unity). The example shown here consists of 20 points (the sample interval equals 0.05 m). Unit impulses produce entirely flat power spectra. The reason is that the unit impulse can only be reconstructed using regular sinusoids from every frequency and with each component having equal amplitude. Note that the component sinusoids are aligned (or have phases) such that all the oscillations cancel each other, except at one point where the contributions sum to unity.

that has a spectrum with approximately equal background power at all frequencies or **white noise** (white light has equal contributions of different frequencies, hence the term pre-whitening). In other words, this type of pre-processing is used to reduce the dynamic range (e.g. Hays *et al.*, 1976; Priestley, 1981).

Pre-whitening enhances the high frequencies and suppresses the low frequencies (Figs. 3.7 and 3.8). Its effect is like a weak form of high-pass filtering (Section 4.3). It is especially useful if the slope of the spectrum is very steep and the frequencies of interest are distant from the left-hand (low-frequency) side of the spectrum. To perform pre-whitening, each successive value is subtracted from the next after multiplying the next value by a weighting factor (Emery and Thomson, 1997). The weighting factor allows for the degree of similarity, or **autocorrelation**, of successive values. The weighting factor equals the correlation of the whole time series with itself after being offset at each point by one value, i.e. the 'lag-1 autocorrelation' or ρ_1 (autocorrelation is discussed further in Section 3.4.4). This can be represented as $xp(t) = x(t) - \rho_1 x(t+1)$, where $x(t)$ represents the time series value at position t and $xp(t)$ represents the corresponding pre-whitened value.

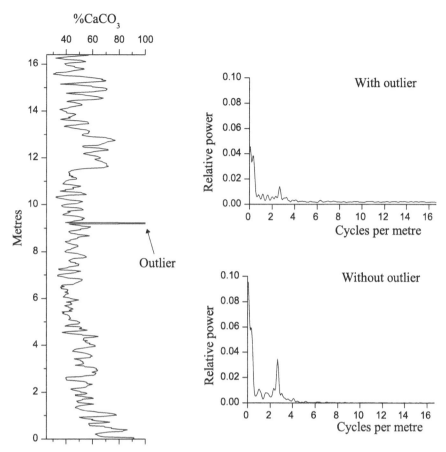

Fig. 3.6 Outliers in time series should be removed prior to time-series analysis. The Belemnite Marls carbonate data in this case consist of 548 data points. However, inserting a single artificial value as an outlier has a profound effect on the spectrum. The flat part of the distorted spectrum corresponds to the spectrum of the unit impulse (see Fig. 3.5) which has raised the level of the high-frequency noise.

In simple pre-whitening the weighting factor equals one so that successive values are just subtracted from each other (a process called '**first differencing**'). Simple pre-whitening tends to suppress the low frequencies too much. This is especially obvious on spectra plotted as log power versus frequency (Figs. 3.7 and 3.8). In the example of the Belemnite Marls carbonate data, the lag-1 autocorrelation (ρ_1) equals 0.917. When this value is used for pre-whitening the result is more satisfactory. However, Mann and Lees (1996) argued that the raw lag-1 autocorrelation is overestimated if there are regular cyclic components in the data. They provided a 'robust method' to estimate the somewhat lower 'true' lag-1 autocorrelation based on fitting a line to the median smoothed spectrum (see Section 3.5). The robust estimate of ρ_1 for the Belemnite Marls data is 0.860 and when used for pre-whitening the resulting background spectrum is much flatter at low frequencies than using the raw ρ_1 estimate (Fig. 3.8).

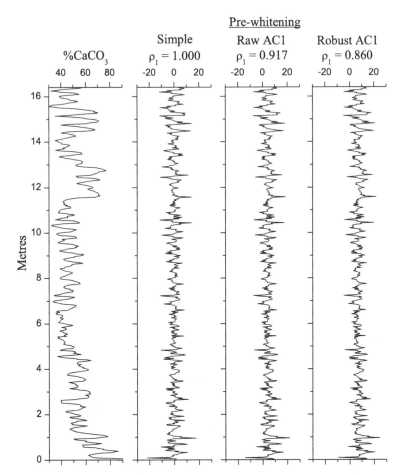

Fig. 3.7 In some time series there is a dominance of low-frequency components which makes it difficult to examine higher frequency spectral peaks. One approach is to pre-whiten the data by modifying the time series so that each new value represents the difference between successive original values (i.e. 'first differencing'). Here several pre-whitening results are shown for the Belemnite Marls using different values of the lag-1 autocorrelation coefficient (ρ_1, AC1, see text for explanation). The corresponding spectra are illustrated in Fig. 3.8.

3.2.5 Other data transformations

Time series with non-stationary variance can sometimes be made more stationary by finding the log of the data values. This is the case when there are some intervals of data with a small mean and small amplitude oscillations (small variance) and other intervals with a large mean and large amplitude oscillations (large variance). By taking the log of the values one can often achieve **variance stabilization**. In other words, the scale of variability after transformation is no longer correlated with the size of the local mean (Fig. 3.9, e.g. Hinnov and Goldhammer, 1991). If the data values range down to zero, logs cannot be used unless a value of one is added to

Pre-whitening

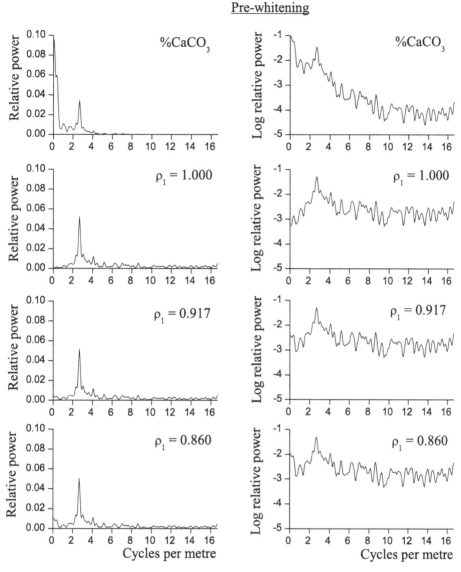

Fig. 3.8 Pre-whitening of the Belemnite Marls data (Fig. 3.7) suppresses the low-frequency spectral peaks while amplifying the high-frequency peaks. Note that just first differencing the original values (using weighting factor of 1.0) has the effect of suppressing the lowest frequencies too much. Using either the raw ρ_1 determined directly from the original time series or a robust estimate produces spectra that are flat in the low-frequency region.

each value first (called a '**log $(x + 1)$ transformation**'). Similarly, negative numbers can be handled by initially adding a positive constant so that all the values become positive. Once the log-transformed data have been detrended, so that the mean is approximately constant, the data can then be considered stationary in terms of both mean and variance. Log transformation will affect the nature of the power

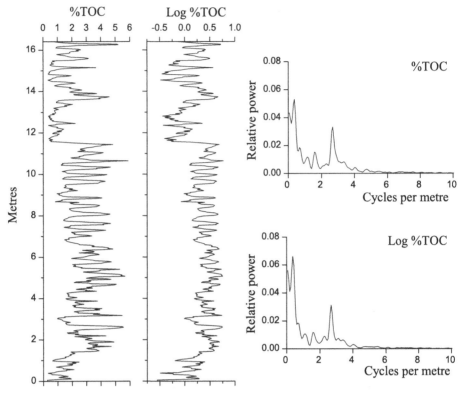

Fig. 3.9 In some time series small amplitude oscillations occur where the average values are small and large amplitude oscillations occur where the average values are large. Alternatively, the data may be skewed with flat troughs and narrow peaks. Taking logs of the values can be used to produce a record with more stable variance in the first case or more sinusoidal oscillations in the second. In the case of the Belemnite Marls total organic carbon data (%TOC), the first case applies, but a log transformation has little impact on the spectrum.

distribution in the power spectrum, but at least the spectral peaks can be considered to be more representative of the average power (variance) of the whole record, rather than being disproportionately influenced by short intervals of large variance (Section 3.2.2).

In other cases non-sinusoidal oscillations of time series derive from the nature of the encoding process and/or from diagenetic factors (Sections 5.2.4 and 5.3.1). For example, cuspate or skewed time series have many low values of a variable (wide troughs) and few high values (narrow peaks), or vice versa. However, when the cause of this skewed shape is understood and analysis is concerned with the presence of the basic cyclicity rather than the particular shape of cycles, it is useful to be able to remove the cuspate shapes. Taking the log of the time series values can sometimes produce a more nearly sinusoidal record (Tong, 1990). When data are collected as proportions (0.0–1.0) or percentages, values near the limits of the range can 'bunch up' rather than exhibit a normal distribution. In this case the so-called **arc sine transformation** can produce more normally distributed values (i.e. use the arc sine of the square root of the values or, for percentages, divide by 100.0 and then square root and arc sine). In the example of the Belemnite Marls carbonate data, the values are close to normally distributed around the mean since there are no values less than 25% or over 95%.

3.3   Spectral estimation – preliminary considerations

3.3.1   Classes of spectra and noise models

From a theoretical point of view, there are four classes of power spectra – line, continuum, discrete and mixed spectra (Priestley, 1981; Percival and Walden, 1993). Time series with line (or discrete) spectra consist of just a few component sinusoids with a spectral peak (line) for each component. Such time series can be described easily using equations involving trigonometric terms (sines and cosines) because the oscillations are perfectly periodic. Continuum spectra result from time series that exhibit some form of irregular fluctuations that occur at all scales of variation (which can be considered to be noise), though not necessarily with the same average amplitude (e.g. red noise). Discrete spectra consist of line components and a white noise background (Percival and Walden, 1993). However, of most interest in cyclostratigraphy are mixed spectra (Priestley, 1981). In time series producing mixed spectra there is a mixture of regular or near regular (periodic or quasi-periodic) oscillations that produce spectral peaks, as well as irregular fluctuations which account for a sloping spectral continuum.

In linear models of time series the continuum can be described as the product of random fluctuations. These fluctuations are considered to result from random or **stochastic processes** in the environment as well as random measurement errors. Stochastic processes can only be described/modelled statistically since theoretically there is no equation controlling the fluctuations. Time series that can be described perfectly using equations are described as the product of **deterministic processes**. There are three main types of spectral continuum noise according to linear models. Firstly, truly random numbers correspond to a succession of values that are entirely independent or uncorrelated with each other. Time series consisting of random numbers have oscillations of all scales with approximately equal average amplitudes and thus produce a flat spectral continuum (i.e. white noise or **all-colour noise**, Fig. 3.10). If the distribution of the values around the mean is normal it is described as Gaussian white noise.

Natural physical systems rarely, if ever, produce white noise time series because of environmental 'inertia'. Other types of noise or time series with continuous spectra (no sharp spectral peaks) can be constructed using white noise (random number) time series. If the noise consists of a random number plus a component, or positive or negative 'weighting', of the previous time series value(s), it is described as the product of an **autoregressive process** (or **AR process**, Priestley, 1981). The weighting of the previous time series values just means that the previous values are multiplied by particular

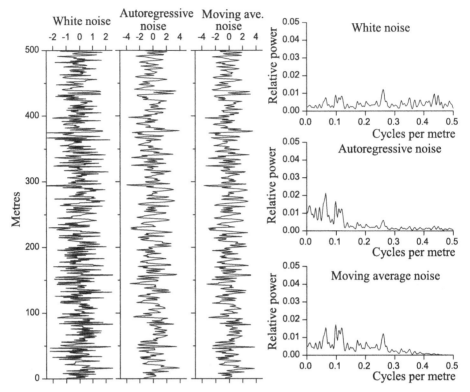

Fig. 3.10 The type of noise encountered in spectra related to environmental cyclicity and cyclostratigraphy can be modelled via linear equations in three main ways. True random numbers (with the mean subtracted and divided by the standard deviation) have successive values that are entirely independent (uncorrelated) with earlier values. As a group the values themselves are distributed normally around the mean, but each value is equally likely to appear in a time series. The result is described as Gaussian white noise.

If each successive value is partly random and partly depends on the previous value, the outcome is a first-order autoregressive (or AR(1)) noise. Symbolically this can be written as: $X_t = a_1 X_{t-1} + a_2 X_{t-2} + \cdots + \varepsilon_t$ (where X_t means the time series value at position t, a_1, a_2, etc. represent positive or negative coefficients and ε_t means the Gaussian white noise value at position t).

Successive values representing the average of groups of random numbers are described as moving-average (MA) noise. Using the same notation this can be represented as: $X_t = a_1 \varepsilon_{t-1} + a_2 \varepsilon_{t-2} \cdots + \varepsilon_t$. The white noise and AR(1) examples illustrated are taken from Priestley (1981). Simple averages of pairs of the Priestley (1981) white noise data were used to create the MA(1) example shown.

Both the autoregressive noise and the MA noise have spectra that slope towards the high frequencies, so the noise is described as red or reddish (as red light is dominated by low frequencies). The AR(1) and MA(1) noise look very similar, but the spectra have different average slopes. The values illustrated are, of course, entirely artificial. However, to create an analogy with the analysis of real geological series, they were treated as though they were collected as real data at one-metre intervals.

coefficients. The 'autoregression' indicates the partial dependence (i.e. correlation) of the new values on previous values. In physical terms, the system's inertia produces 'persistence' or, in other words, a record that is considered to possess a 'memory' of past events. Time series with such a memory are described as having a Markov property. First-order AR processes (AR(1) processes) have a memory lasting one time step; in other words, new values depend on the immediately preceding value only (plus the random component). The smoothness of the resulting time series values means that higher frequency components have smaller average amplitudes than the lower frequencies, so the spectrum slopes down towards the Nyquist frequency (Fig. 3.10). AR(1) time series are described as examples of red noise. Higher-order autoregressive processes have longer memory (i.e. dependence on several earlier values), which has the effect of setting up oscillations in the time series (Schwarzacher, 1975). As a result, the continuum spectrum contains a maximum at one of the non-zero frequencies. This means that a suspected mixed spectrum can be exceedingly difficult to distinguish from the product of a higher-order autoregressive process (Priestley, 1981).

A special form of AR(1) process is where there is a cumulative summation of successive random numbers. In other words, each new value is just the sum of the old *unweighted* value plus a random number. This is termed a **Brownian walk**, which is characteristically non-stationary and possesses a fractal shape (Fig. 3.11, Middleton *et al.*, 1995; Turcotte, 1997). This model of noise appears to correspond very widely to the type of noise encountered in climatic and cyclostratigraphic records (Section 6.2, Mann and Lees, 1996). Brownian walks possess spectra where the log power decreases by a factor of four for a doubling of log frequency. This relationship is obvious when the spectrum is plotted using log-power versus log-frequency, because the power decreases along a straight line with a slope of -2.0 (Fig. 3.11). Such a linear relationship between log power and log frequency is called a **power law** (Turcotte, 1997).

Another type of noise can be generated by using weighted moving averages of previous random numbers from a list plus a new unweighted random number. The result is termed a **moving-average process** (or **MA process**, Fig. 3.10, Priestley, 1981). For first-order MA processes (MA(1) process) a random number is added to a single weighted random number. For a MA(2) process a random number is added to the sum of two weighted random numbers, etc. Geologically, a moving average is generated by processes such as bioturbation and laboratory homogenization of samples when sediments at different stratigraphic levels and different compositions are mixed. Because of the smoothing effect of the mixing, the spectrum becomes steeper or 'reddened' compared to the expected spectrum (Dalfes *et al.*, 1984; Pestiaux and Berger, 1984b). All forms of interpolation have the effect of reddening the spectrum; in a sense the original data become smoothed (Section 5.4.1, Schulz and Stattegger, 1997). More complex noise models include the mixed **autoregressive-moving-average** or **ARMA processes** where there is a combination of dependence on previous values as well as an averaging of random numbers (Priestley, 1981; Tong, 1990). Pardo-Igúzquiza

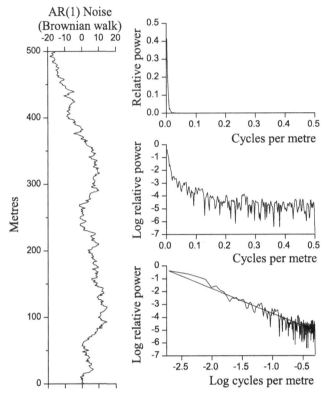

Fig. 3.11 A special form of AR(1) process involves the cumulative sum of unweighted random numbers resulting in a process called a Brownian walk. Using the notation in the caption to Fig. 3.10 this can be represented as: $X_t = X_{t-1} + \varepsilon_t$. Brownian motion, as a model for the motion of microscopic particles suspended in a fluid, occurs in two dimensions so the erratic path can involve reversals in direction. On the other hand, a Brownian walk always proceeds forwards (it is one dimensional), but with variations in the degree to which the path moves to the left or right. The AR(1) time series illustrated represents the cumulative sum of the white noise illustrated in Fig. 3.10.

Notice that this form of an AR(1) process is not stationary (there is a trend present in the data which has to be removed prior to spectral analysis). Brownian walks produce time series where there is a linear relationship between log power and log frequency, i.e. a power law.

et al. (2000) provide computer algorithms for generating time series from various types of stochastic process.

3.3.2   Spectral resolution and bandwidth

Ideally a spectrum can be used to measure the wavelength of regular components by reference to the frequency of the corresponding spectral peaks. However, in practice power spectra have a finite resolution which depends on: (a) the amount of data present in the time series (i.e. the number of data points) and (b) the type of method used to generate the spectrum (i.e. the spectral estimation technique). Thus for a fixed

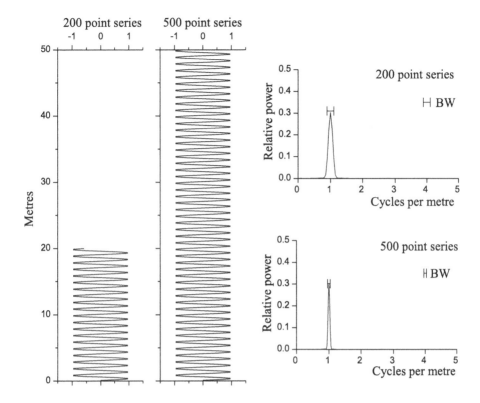

Fig. 3.12 The frequency resolution or bandwidth (BW) of a spectrum is related to the nature of the spectral estimation technique, the sample interval and the amount of data in the time series. For a particular sample interval, the longer the time series, the narrower the bandwidth. The frequency value of every spectral peak has an uncertainty corresponding to $\pm \frac{1}{2}$ BW. In this figure the bandwidths have been shown on top of the spectral peaks to indicate the frequency resolution.

In this example the regular cycle has a frequency of 1.0 cycles per metre and a wavelength of 1.0 m. In the 200 point record the BW = 0.210 cycles per metre ($\frac{1}{2}$ BW = 0.105 cycles per metre) so the uncertainty in peak position ranges from $1.0 - 0.105$ to $1.0 + 0.105$, or 0.895–1.105 cycles per metre. This corresponds to an uncertainty in the wavelength of the regular cycle of 1.12–0.91 m. Similarly, in the 500 point record the BW = 0.084 and the uncertainty in frequency is reduced to 0.958–1.042 cycles per metre and the corresponding wavelength uncertainty is 1.04–0.96 m. Notice that even with 50 repetitions of the regular cycle in the time series, the uncertainty in wavelength in this case is still 8% (i.e. $(1.04/0.96) \times 100\%$).

sampling interval, the more data present in the time series, the higher the frequency resolution (Fig. 3.12). This phenomenon reflects the fact that in spectral analysis it is the data themselves that are used to generate the spectrum. The greater the number of data points, the more spectral information can be packed into the region between the left-hand side of the spectrum (or zero frequency) and the Nyquist frequency. For spectra based on the discrete Fourier transform (Section 3.4.1), the spacing of successive spectral estimates along the frequency axis is defined by the **Rayleigh**

frequency which equals one over the number of data points times the sampling interval ($1/N \times$ SI).

The frequency resolution of a spectrum is described by the **resolution bandwidth** (BW). On the other hand the **signal bandwidth** (frequently referred to in the context of data transmission and the communications industry) indicates the frequency range of the spectrum. It is normally apparent from the context whether the signal bandwidth or resolution bandwidth is being referred to; hence one normally sees the term bandwidth used alone. The definition of resolution bandwidth varies (Bloomfield, 1976; Priestley, 1981), but it is always related to the width of the narrowest detectable spectral peak. Bandwidth is always larger than the Rayleigh frequency. Bandwidth or frequency resolution is constant across a spectrum if the frequency axis is plotted linearly and is usually illustrated using a horizontal bar. The frequency of each spectral peak has an uncertainty of $\pm \frac{1}{2}$ BW (Fig. 3.12). However, if power, or log of power, is plotted against the log of frequency, a single bar cannot be used to illustrate the bandwidth; it should be quoted in a figure caption instead.

The bandwidth of a spectrum has an important consequence in the determination of the wavelength of regular cycles. Consider a spectrum where the bandwidth equals 0.2 cycles per metre and there are two spectral peaks, one at 2.0 cycles per metre and one at 0.2 cycles per metre (Table 3.1). The spectral peak at 2.0 cycles per metre corresponds to oscillations with a 0.50 m wavelength. However, the bandwidth means that this peak could correspond to oscillations up to half a bandwidth either side of the peak frequency. So the peak at 2.0 cycles per metre might actually include oscillations ranging from 2.0–0.1 cycles per metre to 2.0+0.1 cycles per metre or a total range of 1.9 to 2.1 cycles per metre. In other words, oscillations in the time series varying from 0.53 to 0.48 m could be contributing power to the same spectral peak. In the case of the peak at 0.2 cycles per metre the corresponding oscillations could range from 0.1 to 0.3 cycles per metre or a wavelength range of 10.0–3.3 m. In this latter case one could not argue that the spectral peak definitely corresponds to a regular cyclicity because the range of possible wavelengths contributing to the spectral peak is so large. A narrow bandwidth is clearly desirable when trying to detect regular cyclicity. Reducing the bandwidth is only possible by lengthening the time series (rarely an option) or by adopting an appropriate method of spectral estimation where the bandwidth is minimized in some fashion (Sections 3.4 and 5.4.1).

Table 3.1. *The effect of frequency and bandwidth on wavelength uncertainty ($\frac{1}{2}$ BW $= 0.1$ cycles per metre)*

	Peak $- \frac{1}{2}$BW	Spectral peak	Peak $+ \frac{1}{2}$BW
Frequency (cycles per m)	1.9	2.0	2.1
Wavelength	0.53 m	0.50 m	0.48 m
Freq. (cycles per m)	0.1	0.2	0.3
Wavelength (m)	10.0 m	5.0 m	3.3 m

3.3.3   Data tapering, spectral side-lobes and bias

The data in real observational data stop abruptly at either end of the record; this produces discontinuities during a Fourier transformation. The result is that the discontinuities introduce unwanted power into the spectrum that is associated with every 'real' spectral peak. This is illustrated in Figs. 3.13 and 3.14 for the **periodogram**, which can be considered to be a raw form of spectrum. Effectively, some of the power expected in each real peak has escaped laterally, hence the term **periodogram leakage** (Priestley, 1981). This means that the spectral power is incorrectly estimated, so the results are described as exhibiting **bias**. The bias of the periodogram means that power at expected spectral peaks is slightly underestimated and the leaked power raises the spectrum elsewhere (Percival and Walden, 1993). Leakage occurs because in addition to the expected component sinusoids of the series, the step-like end discontinuities can only be represented using multiple additional sine and cosine components. The

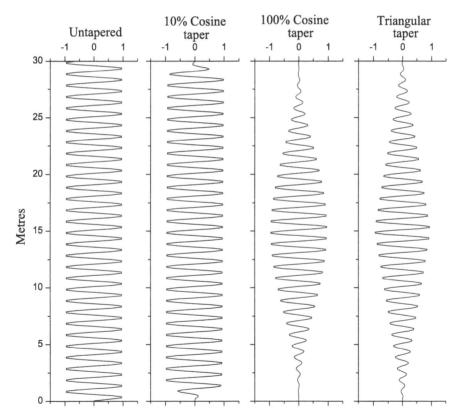

Fig. 3.13 Tapers or data windows applied to a time series have the effect of suppressing the periodogram leakage in ways that depend on the exact form of the taper (Fig. 3.14). Tapering is applied by multiplying the data values by certain coefficients that vary along the length of the time series, usually ranging from zero at the ends to unity at the centre. When a cosine taper is applied to the whole length of the time series (i.e. the 100% cosine taper) the coefficients are also known as the Hanning taper. The triangular taper decreases linearly from unity in the centre of the data to zero at the ends.

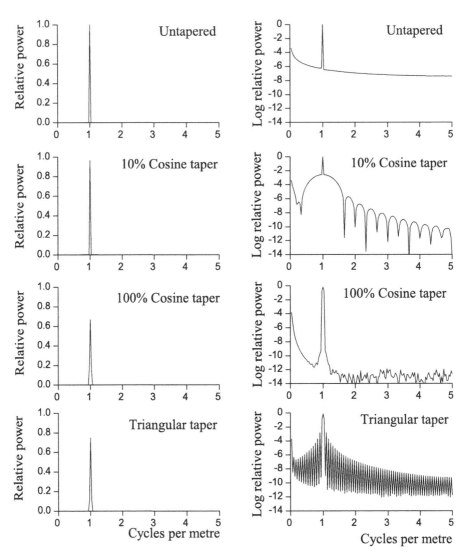

Fig. 3.14 The periodograms of the time series in Fig. 3.13. The log power periodogram of the untapered data (top) shows that periodogram leakage has occurred on either side of the spectral peak. Periodogram leakage arises from the necessarily abrupt ends to observational time series. The abrupt ends to the data are represented in Fourier analysis by small amplitude components from every component frequency. Tapering suppresses periodogram leakage (i.e. reduces the power of the leakage), but introduces spectral side-lobes. Tapering is especially important in time series with red-noise continuum spectra, because otherwise the small high-frequency spectral peaks can be masked by power from periodogram leakage. The greater the proportion of the time series that is tapered, the greater the suppression of periodogram leakage, but the wider the main spectral peaks (spectral smearing).

discontinuities are imperfectly represented by allocating a small but varying amount of power from every frequency in the spectrum. This is reminiscent of the representation of the unit impulse where every frequency contributes some power (Fig. 3.5). The imperfect representation of the end discontinuities, from the addition of multiple sinusoids, is called the **Gibbs' phenomenon** and can cause ringing (Proakis and Menolakis, 1996).

Only a small amount of power leaks away from the expected periodogram peaks. However, the red noise characteristics of cyclostratigraphic data (a large dynamic range) mean that periodogram leakage from low frequencies can swamp small-amplitude peaks at high frequencies. Periodogram leakage can be partly alleviated by making the oscillations die away gradually at each end of the time series (Fig. 3.13). This procedure is known as **tapering**, applying a **fader** or **data windowing**. Tapering consists of multiplying the time series by a series of values, or window weights, that start at one in the centre of the data (i.e. no tapering) and drop to zero at the edges. However, the tapering produces spectral **side-lobes** or minor peaks associated with each real spectral peak. The nature of the side-lobes depends on the shape of the taper and the proportion of data that are tapered (Figs. 3.13 and 3.14, Priestley, 1981; Ifeachor and Jervis, 1993). Tapering can suppress periodogram leakage sufficiently to allow detection of small-amplitude, high-frequency peaks. Notice that when tapering is applied to a large proportion of the time series the real spectral peaks become wider (Fig. 3.14). This is known as **spectral smearing** and represents an undesirable feature of tapering.

3.3.4   Comparing spectra and normalization

It is often the case in cyclostratigraphy that the absolute power values in a spectrum are unimportant. What usually matters is the shape of the spectrum. However, different time series have different average amplitudes and this means that the absolute power values vary considerably between records. This is awkward when comparing spectra of different variables obtained from the same strata. Some authors leave off the scale values on the power axis altogether. An alternative approach is to re-scale the spectra so that the maximum power value in each spectrum is given a value of one. A generally more satisfactory procedure is to divide all the power values by the sum of the power values so that the relative power is obtained (Fig. 3.15). Alternatively this normalization can also use division by the average power. Yet another procedure is to standardize the time series prior to spectral estimation (i.e. divide each detrended time series value by the standard deviation). None of these methods changes the shape of the spectrum.

As well as normalizing the power range to allow comparison of spectra, it is common for mathematical texts to standardize the frequency range. Often a standardized sampling interval (i.e. SI = 1.0) is used, so that all the spectra illustrated have frequency axes running from 0 to 0.5. **Angular frequency** (denoted ω) is measured in radians and is simply the frequency times 2π. With a standardized sample interval this yields spectra with frequency axes running from 0 to π.

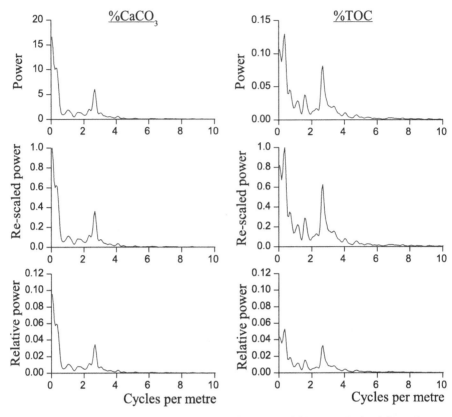

Fig. 3.15 When time series for two different variables are obtained from the same samples the range of values differ and thus so do the power values, even if the regular cyclicity is almost identical. This figure shows the spectra relating to the Belemnite Marls carbonate (30–90%) and organic carbon (0–6%) data from the basal 16.41 m of the unit (see Fig. 1.10). In order to compare spectra for different variables it is often convenient to be able to plot the spectra using similar power axes. One approach is to plot log power values versus frequency. For linear plots of power versus frequency there are two methods available. Each power value can be divided by the maximum power value so that the largest power value equals one (middle spectra). Alternatively, the power values are divided by the sum of the power values to yield the relative power (bottom spectra). Neither method changes the shape of the spectrum, but using relative power is usually more satisfactory when the maximum power occurs at different frequencies in the two spectra (as seen here).

3.4   Spectral estimation – methods

3.4.1   The Fourier transform and the periodogram

The periodogram was introduced as a method specifically for estimating line spectra (Schuster, 1898; Priestley, 1981). It performs badly with time series that possess a continuum spectrum unless the background corresponds to white noise (Percival and Walden, 1993). Due to the prevalence of red noise in cyclostratigraphic records (Section 6.1) the periodogram should not be used for spectral analysis in contexts such as those described in this book. Nonetheless, it forms the basis for many techniques used to calculate spectra (spectral estimation). The periodogram is obtained by use of the **discrete Fourier transform** or **DFT**. The Fourier transform manipulates a time series in a manner that does *not* depend on the nature of the data. Hence the transform is not a form of modelling and spectral estimation techniques based on the Fourier transform are therefore sometimes called non-parametric (Ifeachor and Jervis, 1993).

The way in which the DFT operates is illustrated in Figs. 3.16a and 3.16b. The importance of each different frequency component in the time series is ascertained by using paired sine and cosine waves. Note that in the formula for the Fourier transform, many mathematical texts compress the separate description of each sine and cosine component into a single complex number (the real part representing the sine component and the imaginary part the cosine component). It is common for such texts to use angular frequency (ω, Section 3.3.4) rather than frequency (f) in their formulae for the Fourier transform.

During the discrete Fourier transform, the time series is multiplied, point-by-point, by a cosine wave of a particular frequency. The results are added up and multiplied by a certain constant ($2/N$ where $N =$ the number of data points). This yields the average amplitude of that frequency cosine component. (The calculations proceed as though half the spectrum exists as a mirror image of the real spectrum at 'negative frequencies'. Since negative frequencies have no physical meaning for observational time series, the average amplitude has to be doubled using the constant.) Next the time series is multiplied by a sine wave of the same frequency and again the results are summed and multiplied by the constant (Figs. 3.16a and 3.16b). For each frequency component examined, the relative size of the sine and cosine average amplitude dictates the average phase of the time series oscillations (Fig. 1.2).

The Fourier transform can be thought of as re-organizing the time series data into a different arrangement based on frequency. If the result is not modified (e.g. power is not calculated), the **inverse Fourier transform** can be used to re-create the original time series. This interchangeability via the Fourier transform has led to the concept of time series data present either in the **time domain** or the **frequency domain**.

The sum of the squared sine and cosine average amplitudes yields the periodogram power at the frequency of interest. The plot of power (or squared average amplitude) versus the frequency yields the periodogram. Figures 3.16a and 3.16b show that the number of spectral values in the periodogram is simply one plus half the number of time series points.

The discrete Fourier transform has a number of drawbacks. Firstly, it is slow computationally so a group of rapid computer algorithms was developed to perform the **fast Fourier transform** or **FFT** (Press *et al.*, 1992). The FFT algorithms usually require data sets that have an integer power of two points (e.g. 256, 512, 1024 etc.). If necessary, zeroes can be added to the data to fulfil this constraint, but the effect of this '**zero-padding**' needs to be compensated for when calculating bandwidths and confidence intervals (Section 5.4.1, Bloomfield, 1976). Despite the fact that the DFT is

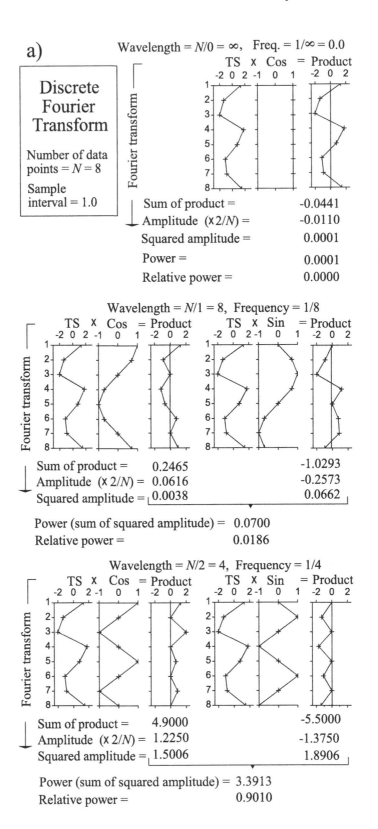

a) Wavelength = $N/0 = \infty$, Freq. = $1/\infty = 0.0$

Discrete Fourier Transform

Number of data points = $N = 8$

Sample interval = 1.0

Sum of product =   -0.0441
Amplitude $(\times 2/N)$ =   -0.0110
Squared amplitude =   0.0001
Power =   0.0001
Relative power =   0.0000

Wavelength = $N/1 = 8$, Frequency = 1/8

Sum of product =   0.2465   -1.0293
Amplitude $(\times 2/N)$ = 0.0616   -0.2573
Squared amplitude = 0.0038   0.0662

Power (sum of squared amplitude) = 0.0700
Relative power =   0.0186

Wavelength = $N/2 = 4$, Frequency = 1/4

Sum of product =   4.9000   -5.5000
Amplitude $(\times 2/N)$ = 1.2250   -1.3750
Squared amplitude = 1.5006   1.8906

Power (sum of squared amplitude) = 3.3913
Relative power =   0.9010

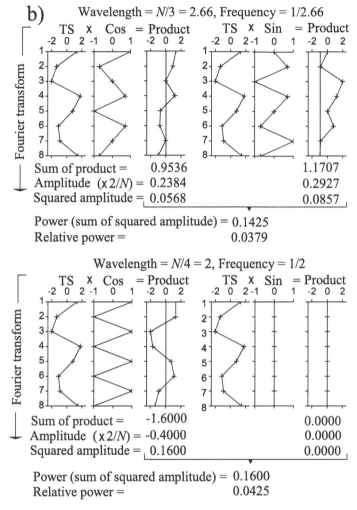

b)   Wavelength = $N/3$ = 2.66, Frequency = 1/2.66

Fourier transform

TS × Cos = Product		TS × Sin = Product	

Sum of product =   0.9536   1.1707
Amplitude ($\times 2/N$) = 0.2384   0.2927
Squared amplitude = 0.0568   0.0857

Power (sum of squared amplitude) = 0.1425
Relative power =   0.0379

Wavelength = $N/4$ = 2, Frequency = 1/2

Fourier transform

TS × Cos = Product		TS × Sin = Product	

Sum of product =   -1.6000   0.0000
Amplitude ($\times 2/N$) = -0.4000   0.0000
Squared amplitude = 0.1600   0.0000

Power (sum of squared amplitude) = 0.1600
Relative power =   0.0425

Fig. 3.16 a,b) The Fourier transform decomposes time series into component regular sines and cosines. This involves multiplying the time series values by cosine and sine waves of different frequencies. If the discrete Fourier transform is used (including the fast Fourier transform), the frequencies examined are defined by integer divisors of the number of time series points, N, times the sample interval, SI, or $1/(N \times SI/0)$, $1/(N \times SI/1)$, $1/(N \times SI/2)$, $1/(N \times SI/3)$, ...). The highest frequency component is always $1/(N \times SI/(N/2))$ or $1/(2 \times SI)$, the Nyquist frequency (see Fig. 2.6). For each frequency considered, the sum of the cosine and sine products, times two and divided by N, yields the average amplitude.

In this example, the time series (TS) consists of a regular cycle with a wavelength of four units with an initial phase of 35° plus very small-amplitude white noise. Since the time series is dominated by oscillations with a frequency of 1/4, the corresponding sums of the cosine and sine products, and thus the power value, are much larger than for the other frequencies.

slow, modern PCs are so fast that the time for computation is not important except for time series that are thousands of points long, or when, as for electronic engineering, the results are needed almost instantly.

Secondly, the DFT (including the FFT) can generate power values at only certain fixed frequencies (Figs. 3.16a and 3.16b). Using $N =$ the number of time series values and SI = the sample interval, these frequencies occur at: 0.0, $1/(N \times SI)$, $1/(N \times SI/2)$, $1/(N \times SI/3)$, $1/(N \times SI/4)$, ... $1/((N \times SI)/(N/2))$. The last value can also be represented as $1/(2 \times SI)$ and corresponds, of course, to the Nyquist frequency. Since these frequencies are related to each other by the inverse of integers they are described as **Fourier harmonic frequencies**. Sometimes the periodogram is labelled with **harmonic number** (the integers) rather than the actual frequency values. In passing, it is worth mentioning that the phrase **harmonic analysis** refers to analysis of time-series analysis in terms of precisely specified (usually known) wavelength components (Percival and Walden, 1993).

Periodogram power values are independent, so no inference can be made about the power between the harmonic frequencies (for example by interpolation). To reflect this, some authors plot periodograms with vertical lines or unconnected dots for each spectral estimate. If a regular cycle in the data has a wavelength that is not an exact integer divisor of the length of the record, the corresponding peak occurs, slightly reduced in size compared to expectation, at the adjacent frequencies where periodogram power has been calculated. This effect is called **scalloping loss** or **picket fencing** (Ifeachor and Jervis, 1993). With the FFT it is feasible to use zero-padding to substantially increase the length of the time series (sometimes by several times the number of data points) so that the frequency resolution is increased (e.g. Muller and Macdonald, 1997a, b). This is only really necessary if there is reason to suspect the presence of two regular cycles with very similar frequencies, and high-resolution spectral techniques (discussed later) are inappropriate or unavailable.

When a periodogram is used to examine an observational time series, even with prior tapering of the time series, the result is a rather poor approximation to the expected spectrum. This is because the periodogram power values are randomly distributed above and below the expected continuum spectrum (Priestley, 1981; Ifeachor and Jervis, 1993). Different parts of a stationary time series yield different erratic periodogram fluctuations. Increasing the amount of data in the time series does not result in a smoother periodogram; it just increases the number of periodogram values and frequencies (thereby reducing the bandwidth). Thus the periodogram is described mathematically as '**inconsistent**'. Ideally the periodogram needs to be smoothed to reduce the erratic fluctuations and thus better approximate the expected spectrum. Generally, any form of smoothed periodogram, or the output from a method of spectral estimation which approaches the theoretical spectrum better than a periodogram, is known as a power spectrum.

Effectively each periodogram value is based on two values or two **degrees of freedom** (i.e. the squared mean cosine amplitude plus the squared mean sine amplitude).

Most methods that result in smoothed periodograms have the effect of increasing the degrees of freedom. This means that the result is closer to the expected spectrum because more data are used in the calculation of individual spectral estimates. This extra smoothing is often obtained by using information from adjacent frequencies; the result is a better approximation of the continuum part of the spectrum.

However, the smoothing increases the bandwidth, making it harder to identify the line components that are needed to identify regular cyclicity. This means that a compromise between smoothness and spectral resolution needs to be reached for the sorts of mixed spectra encountered with cyclostratigraphic data. The choice of the degree of smoothing, and thus the degrees of freedom, is highly subjective, though extensively discussed in the literature (e.g. Priestley, 1981). In cyclostratigraphic studies typically between 8 and 14 degrees of freedom are used (except for the maximum entropy method). For many spectral windows the degrees of freedom equals two divided by the variance of the spectral window (Bloomfield, 1976). Note, however, that the variance of the spectral window depends on the type of window as well as correction factors for the proportion and type of data tapering and, if used, the amount of zero-padding of the time series.

3.4.2   The direct method

One method used to smooth the periodogram is to use a moving average applied to the periodogram values. The particular moving average applied is termed a **discrete spectral window** or **frequency window** and each window type is often named after its author or authors. Discrete spectral windows should always be used with data tapering in order to suppress periodogram leakage.

The simplest moving average is unweighted and called **box-car smoothing** or the use of the Daniell spectral window. This simply finds the average of all the periodogram values occurring within the window. The wider the window, the smoother the resulting spectrum. Multiple applications of a short Daniell window also produce a smoother spectrum, but the results are more satisfactory than using a single application of a wide window because otherwise the spectral peaks become rather flat-topped (Bloomfield, 1976).

Other types of discrete spectral windows use weighted moving averages. For example, applying the three-point weight sequence 0.25, 0.5, 0.25 (and end-point weights 0.5, 0.5) to the periodogram values is termed **Hanning** and these weights are referred to as the Tukey or Tukey–Hanning discrete spectral window (Fig. 3.17, Priestley, 1981). In contrast to the use of one application of the Daniell window, the weighted moving-average spectral windows have the effect of producing rounded spectral peaks. This procedure does not produce side-lobes and in fact helps suppress the side-lobes that result from tapering (Figs. 3.14 and 3.17). Repeated application of a weighted moving average can be used to smooth the spectrum as much as desired (Fig. 3.18). Spectral confidence intervals (Section 3.5) and bandwidth are determined using the variance of spectral estimates. For discrete spectral windows, the variance of the spectral estimates

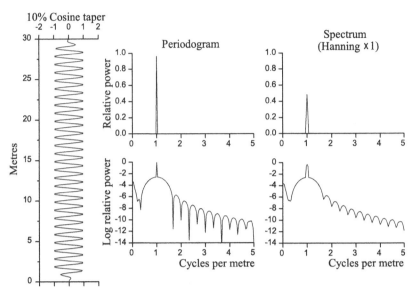

Fig. 3.17 For time series with some form of spectral continuum (e.g. a red noise background) it is very important to be able to reduce the erratic fluctuations in the periodogram values. One form of discrete spectral window consists of a moving average applied to the periodogram values with weights 0.25, 0.5, 0.25 (known as Hanning). The effect is to smooth the periodogram and reduce the amplitude of the side-lobe oscillations (increasing the degrees of freedom from 2.0 to 5.3, or 5.0 allowing for tapering).

is calculated as the sum of the squares of the moving-average weights allowing for the number of moving-average applications, data tapering and zero-padding (Bloomfield, 1976).

An alternative way to produce a spectrum is to average periodograms from equal length subsections of the data after each subsection has been detrended and tapered. If the subsections are consecutive this is termed the **Bartlett method**, and, if they overlap, the **Welch method** (Fig. 3.19, Ifeachor and Jervis, 1993). The degree of smoothing is determined by the amount of overlap of data subsections and the number of subsections used. For a given number of degrees of freedom the Welch method produces better 'quality' spectra compared to the use of a discrete spectral window or the Bartlett method (Ifeachor and Jervis, 1993). Optimum results can be achieved using 50% overlap of the subsections (Percival and Walden, 1993). However, by using subsections with fewer data points than the whole record, the number of frequency values is reduced accordingly (i.e. the resolution bandwidth is increased).

3.4.3 The multi-taper method

Thomson (1982, 1990) advocated an alternative approach to spectral estimation based on the periodogram. This has become known as the **multi-taper method** (MTM)

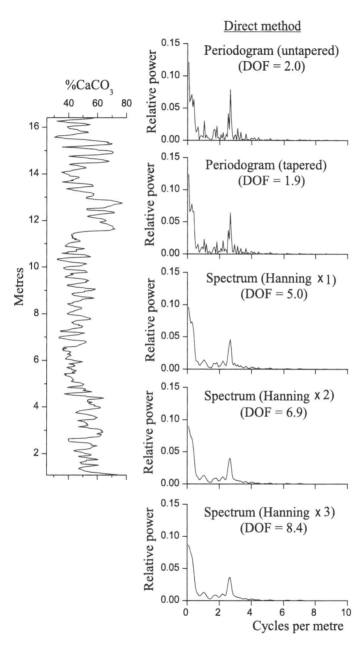

Fig. 3.18 This figure compares the periodograms of untapered and tapered, detrended Belemnite Marls data, with power spectra derived from the periodogram with various degrees of smoothing using a discrete spectral window (i.e. using a weighted moving average). In each case the degrees of freedom (DOF) are indicated. Note that eight degrees of freedom is normally the minimum considered sufficient in cyclostratigraphic studies.

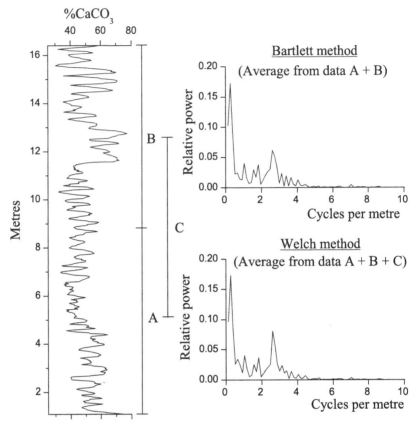

Fig. 3.19 The Bartlett and Welch methods of spectral estimation use averages of the periodograms of detrended and tapered subsections of the time series (labelled A, B and C here). In both procedures the frequency resolution is less than for other spectral methods due to the fewer data points involved for each component periodogram (compare with Fig. 3.18). Greater smoothing can be achieved for the Bartlett method by using more, shorter, consecutive data subsections than shown here. However, this leads to still lower frequency resolution. In the Welch method greater smoothing can be achieved by using more subsections that are half the length of the complete record, but overlapping them by more than the 50% shown in the figure.

or **Thomson tapering**. A series of special data windows, known as **discrete prolate spheroidal sequences** or **Slepian sequences**, are used to taper the time series (Fig. 3.20). Discrete prolate tapers are particularly efficient at suppressing periodogram leakage (Percival and Walden, 1993). The number of tapers used generally ranges from four to eight. After each taper has been applied to the data a periodogram is generated. Since the tapers are orthogonal, each periodogram is based on different weighting for different parts of the data and hence, in the simplest form of MTM, the periodograms can be averaged (Percival and Walden, 1993; Yiou *et al.*, 1996). The effect is to produce a spectrum with well-suppressed side-lobes (i.e. small bias), good smoothing and yet high frequency resolution. However, there is a tendency to produce flat-topped spectral peaks.

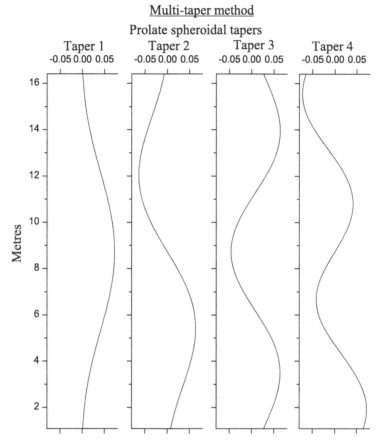

Multi-taper method
Prolate spheroidal tapers

Taper 1	Taper 2	Taper 3	Taper 4
-0.05 0.00 0.05	-0.05 0.00 0.05	-0.05 0.00 0.05	-0.05 0.00 0.05

Fig. 3.20 In the multi-taper method (MTM) of spectral analysis a series of prolate spheroidal tapers are applied to the time series (compare with Fig. 3.13). The different tapers suppress different parts of the time series. The spectrum is then based on the average of the periodograms generated from each tapered series. The number of tapers used can be varied, though the shapes of the tapers vary accordingly. The diagram shows the example of four tapers used for point-by-point multiplication with the Belemnite Marls data in Fig. 3.21. Note that unlike tapers used in the direct method of spectral estimation, the tapers used for MTM spectra take on negative as well as positive values.

The multi-taper method allows an increase in the degrees of freedom compared to the periodogram without the same rate of widening of bandwidth encountered with the other main methods of spectral analysis. The degrees of freedom are approximately twice the number of tapers (Percival and Walden, 1993) though allowance needs to be made for zero-padding (Bloomfield, 1976). The bandwidth is just the Rayleigh frequency times the number of tapers. Computationally the method is complex, but an F-test (Williams, 1984) can be used to test for the presence of spectral peaks related to regular cyclic components with constant phase (Thomson, 1990). Park and Herbert (1987), Olsen and Kent (1996) and Mann and Lees (1996) provide examples of the multi-taper method applied to cyclostratigraphic data. The

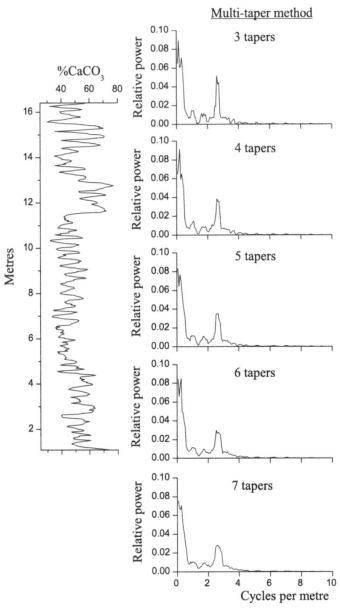

Fig. 3.21 The multi-taper method of spectral estimation for the Belemnite Marls data. Unlike the Bartlett or Welch methods, the averaged periodograms are calculated using the whole length of the time series. Each periodogram is calculated using the data once modified by a particular prolate spheroid taper (illustrated for the Belemnite Marls data using four tapers in Fig. 3.20). Greater smoothing (higher degrees of freedom) results from using a greater number of tapers and hence a greater number of averaged periodograms. In this case there is no zero-padding of the data and the degrees of freedom can be taken to be twice the number of tapers.

method does have its detractors (Priestley, 1981; Muller and Macdonald, 1997b). In particular, for spectra with strong spectral peaks and a very small continuum component, higher frequency resolution can be achieved using the maximum entropy method. Additionally, the number of tapers used depends on the choice of the investigator (Fig. 3.21, Percival and Walden, 1993; Yiou *et al.*, 1996). Nevertheless, the multi-taper method represents a good method for producing spectral estimates with high frequency resolution for given degrees of freedom, low bias and a distribution amenable to the location of confidence levels (Mann and Lees, 1996; Section 3.5).

3.4.4   The Blackman–Tukey method

Prior to the widespread use of fast computers, the standard method of spectral estimation was based on the use of the sample **autocovariance** sequence (i.e. the discrete form of the autocovariance function). Autocovariance is determined by comparing a time series with itself once it has been offset by various amounts. The amount of offset is called the **lag**. The lag runs from zero (no offset) to the number of time series points minus one. When the time series has been centred (i.e. the mean subtracted from each point or any linear trend removed) the autocovariance is easily determined. For each lag, or time series offset, the autocovariance is simply the sum of the time series values each multiplied by the offset counterpart, with the result divided by the number of data values. To understand what the autocovariance represents it is worth comparing it to autocorrelation. The sample autocorrelation sequence is just the autocovariance values divided by a constant, the time series variance (for centred data the variance is the sum of the squared time series values, divided by the number of values). See Percival and Walden (1993) for a discussion of these definitions of autocovariance and autocorrelation. As the name suggests, the autocorrelation measures the similarity of the time series with the offset counterpart. The autocorrelation can vary from $+1.0$ to -1.0. Schwarzacher (1975) and Davis (1986) illustrate examples of autocorrelation plots or **correlograms** for different types of time series. For cyclostratigraphic data the autocorrelation always starts at $+1.0$ for zero lag and then generally oscillates between positive and negative values at larger lags (e.g. see the correlogram of the Belemnite Marls data in Fig. 3.22).

The autocovariance sequence contains the same frequency oscillations as the time series, but the phases have been altered (Priestley, 1981). At large lags, the autocovariance is determined from few overlapping time series values (i.e. a small proportion of the total number of values) so the results can be rather noisy. According to the **Wiener–Khintchine theorem** the periodogram of a time series can be obtained by using the Fourier transform of the autocovariance sequence. In the **Blackman–Tukey method** (BTM) the transform of the autocovariance is only used for results up to a certain lag value, the **truncation point** or M. Truncation eliminates the highest lag autocovariance terms (i.e. the most noisy values) prior to the Fourier transform. The

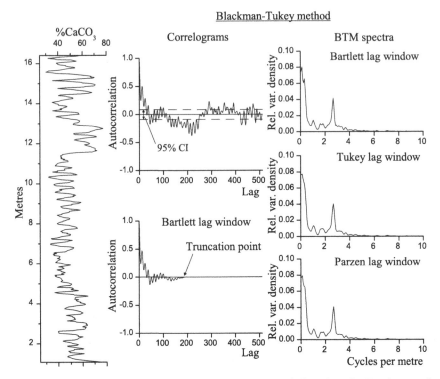

Fig. 3.22 The Blackman–Tukey method (BTM) is based on the Fourier transform of part of the autocovariance sequence. The autocovariance represents the autocorrelation sequence multiplied by the time series variance. Autocorrelation is calculated by comparing the time series with itself with varying numbers of offset steps. The number of offsets is termed the lag. A plot of autocorrelation versus lag is known as a correlogram. The 95% confidence interval for zero correlation equals $\pm 1.96/(N^{0.5})$ and has been shown using horizontal dashes (Williams, 1997).

In the BTM a lag window, or form of taper, is applied to the autocovariance sequence prior to applying the Fourier transform. The largest lag value that is retained, after applying the lag window, is called the truncation point or M. The use of the triangular window illustrated here (the Bartlett window) uses unit weight at zero lag with the weights applied to the autocovariance sequence decreasing linearly towards zero at the truncation point.

The Blackman–Tukey spectra for the Belemnite Marls data are illustrated for three types of lag window. The spectra are very similar for each window type; of more importance is the choice of M. For this figure the value of the truncation point was varied so that for all the spectra shown, though each used a different spectral window, there are eight degrees of freedom.

part of the autocovariance that is retained lies within a '**lag window**'. By using a restricted range of the autocovariance sequence for the Fourier transform, the output is less erratic than the periodogram.

However, if the truncations or lag window edges are abrupt, then large spectral side-lobes result (**spectral leakage**). On the other hand, data tapering is not required in the BTM. To minimize the size of the spectral side-lobes the remaining autocovariance values are weighted from a maximum value at zero lag down to zero at the truncation point M (Fig. 3.22). The position of the truncation point controls the degree of smoothing and hence the degrees of freedom (Priestley, 1981). A larger value of M implies the use of more covariance values and hence a higher-resolution (but more noisy) spectrum. Each form of weighting of the autocovariance values produces a particular form of lag window. Every type of lag window has an associated spectral window with a particular width of main spectral peak and a certain distribution of side-lobes (Priestley, 1981). Lag and spectral windows are named after their authors. Despite the details of different types of lag window/spectral window (Priestley, 1981) there are few differences between the final Blackman–Tukey spectra (Fig. 3.22). Since in this method the power spectrum is obtained using the Fourier transform of the truncated autocovariance sequence, some authors use the term **autospectrum** (e.g. Schulz and Stattegger, 1997).

An advantage of the Blackman–Tukey method is that the frequencies at which spectral estimates are obtained do not need to be restricted to the traditional Fourier harmonic frequencies. Additionally, for the same degrees of freedom, the peaks on BTM spectra are slightly narrower, less erratic and have less bias than for spectra generated via the smoothed periodogram (Ifeachor and Jervis, 1993). Another difference from the spectral methods based on the smoothed periodogram is that in BTM spectra the amount of power associated with each peak is determined by the area under the curve rather than from peak heights. In theory, BTM spectra can contain negative variance density (or power density) values, but this is rarely a problem in work on cyclostratigraphic records.

3.4.5   The maximum entropy method

The other methods of spectral estimation discussed in this chapter require no assumptions about the nature of the data so they are termed 'non-parametric'. However, in the parametric approach it is assumed that the time series can be represented satisfactorily as the result of an autoregressive process of a certain order (Section 3.3.1). There are several approaches to parametric spectral analysis, the most important being the **maximum entropy method** (MEM). In information theory a time series is considered to contain more information when it is less regular. In other words, more information is needed to completely describe an irregular series. For example, data composed of a single regular cycle can be completely described by a single simple equation, and therefore the time series 'contains' little information. The information per unit time is termed the entropy (just as energy per unit time is called power). So entropy measures disorder. The mathematics involves initially calculating the autocovariance sequence from the data out to a certain lag, M. The autocovariance for greater lags is then calculated on the assumption that the time series is as disordered as possible, but with the constraint that the autocovariance implied for 0 to M lags agrees with the results calculated earlier. In other words, the entropy is maximized (Ulrych and Bishop, 1975). This differs from the Blackman–Tukey method, since in that technique it is assumed that the autocovariance is zero beyond M.

MEM is equivalent to fitting the data as though they correspond to a high-order autoregressive (AR) process (Percival and Walden, 1993). The order of the AR process is determined by the number of lags used to determine the autocovariance sequence from the data (i.e. M). Not surprisingly, the choice of M is critical. When complex numbers are used in the explanation of the MEM it is called the **all poles method**, where M equals the number of poles (Press *et al.*, 1992). Maximum entropy spectra do not require prior data tapering and there is no scalloping loss, spectral smearing, or spectral leakage.

There are two main procedures for maximum entropy analysis, known as the Yule–Walker and Burg algorithms. The Yule–Walker algorithm yields accurate spectral estimates, but the spectrum has low frequency resolution compared to the Burg algorithm (Priestley, 1981). Additionally the Yule–Walker estimates exhibit considerable bias compared to the Burg estimates (Percival and Walden, 1993). The Burg algorithm produces rather inaccurate estimates of spectral peak heights, but the spectrum has very high frequency resolution (approximately twice the resolution of the periodogram, Ulrych and Bishop, 1975). The high resolution of the Burg algorithm has been popular in cyclostratigraphic studies. The other approaches to parametric spectral estimation are called the maximum likelihood and least-squares methods (Percival and Walden, 1993).

Like the BTM, the power associated with spectral peaks is found from the area under the curve. The degrees of freedom are N/M (Ulrych and Bishop, 1975). Because the data have been fitted with an autoregressive model, there is no red continuum spectral background in MEM spectra. Naturally, when the continuum spectrum is more appropriately modelled with a moving average than an autoregressive process, the MEM can produce very unreliable spectra (Priestley, 1981). Typically M is chosen between $N/3$ and $N/20$. When M is chosen to be larger than $N/3$, spurious peaks can appear ('**spectral line splitting**', Fig. 3.23, Press *et al.*, 1992). As a result it is prudent to use a non-parametric method of spectral analysis first to establish the number and position of reliable spectral peaks as a guide for the choice of M in the MEM spectrum (Pestiaux and Berger, 1984a). There are routines for selecting M objectively independently of non-parametric spectra (using one of the Akaike criteria, Percival and Walden, 1993). Another problem with the parametric methods, including the Burg algorithm, is that spectral peaks sometimes shift slightly in frequency according to the phase of the regular cycle (Percival and Walden, 1993). Indeed it is worth pointing out that some authors are suspicious of the maximum entropy method, though it is used routinely in digital signal processing (compare Thomson, 1990, with Ifeachor and Jervis, 1993). Aside from the potential to generate spurious peaks, an important objection to the use of the maximum entropy method is that although confidence levels can now be determined, it is only with great difficulty (Percival and Walden, 1993).

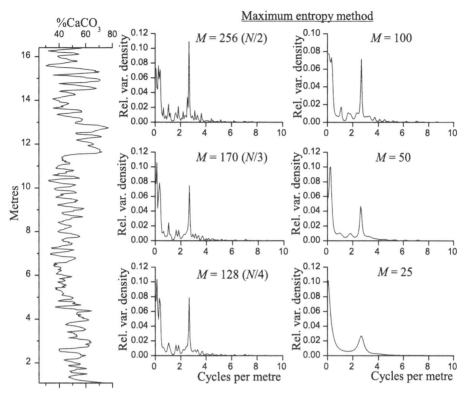

Fig. 3.23 In the maximum entropy method (MEM) the choice of M, which determines the level of smoothing, is critical. With large values of M, fairly large, but spurious, peaks can appear. A value of $M = N/3$ is usually considered the largest acceptable. Note that the heights of maximum entropy peaks obtained using the Burg algorithm are often considered to be unreliable. However, the peak positions are very accurate and the spectral resolution is superior to most other methods of spectral estimation. In the light of this, many authors use plots with log power only and employ the MEM only once the presence of reliable spectral peaks has been established with alternative methods. As for the Blackman–Tukey method, the total power associated with a regular cycle is found by integrating the area under the spectral peak, rather than from peak height.

3.4.6 The Walsh method

Walsh power spectra are based on the use of the square wave orthogonal Walsh SAL and CAL functions instead of sine and cosine waves (Fig. 3.24, Beauchamp, 1984). The equivalent of a periodogram can be generated using the **fast Walsh transform** and, like the FFT, this requires that the number of points in the time series is an integer power of two. Because the functions used have discontinuities, the spectrum is unaffected by the abrupt ends of observational time series; there is no Gibbs' phenomenon. Consequently, the time series does not need to be tapered prior to analysis.

Since Walsh functions have a square wave shape they take on just two values (e.g. +1 and −1). In order to have pairs of functions that account for different phases in a time

Walsh functions

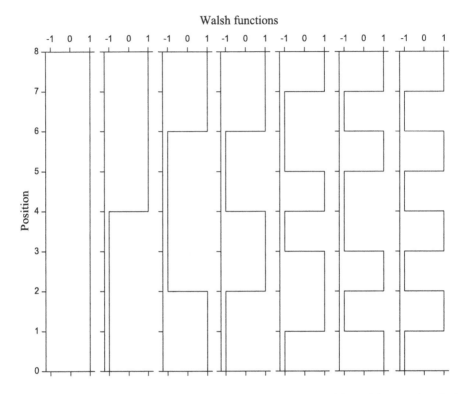

Fig. 3.24 Walsh functions are orthogonal square waves used for Walsh spectral analysis and filtering. Note that some of the functions have variable wavelength oscillations so that data with different phases can be described. This leads to the idea of 'sequency' rather than frequency. Sequency indicates the *average* number of zero-crossings per unit thickness/time. The functions illustrated are arranged with sequency increasing to the right.

series (i.e. the analogue of sine and cosine functions), Walsh functions have variably spaced zero-crossing positions. This means that the wavelengths of the component SAL and CAL components vary along the series (Fig. 3.24). Thus in Walsh analysis one speaks of '**sequency**', meaning average numbers of zero crossings, rather than frequency. The distinction is often ignored in practical work. However, this difference does matter in one respect: the Walsh spectrum changes according to the phase of the time series. To avoid this problem the Walsh periodogram is calculated repeatedly after one time series value has been taken from one end of the data and moved to the other end. This procedure effectively produces periodograms for the complete range of possible phases of the time series. All the periodograms are then averaged and smoothed with a discrete spectral window to produce the **average Walsh spectrum** (Beauchamp, 1984).

As the analysing functions are discontinuous it makes sense to apply this method to time series such as digitized rock type data (Fig. 3.25, Weedon, 1989). If there are just two alternating lithologies these can simply be coded +1 and −1. Where more

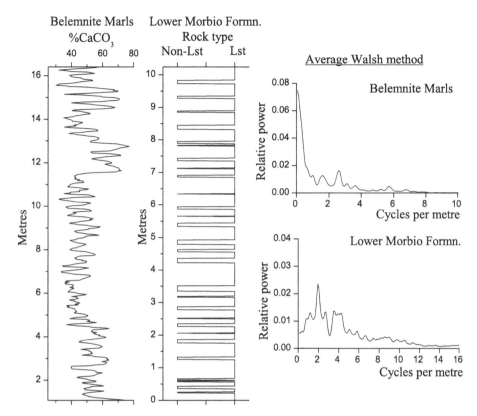

Fig. 3.25 The average Walsh spectrum is based on the use of the square wave Walsh functions rather than sine and cosine waves. This is most suited to examining square wave data such as digitized lithological logs. As an example, data from the Lower Jurassic Morbio Formation of Switzerland are illustrated (Fig. 1.8, Table 1.1, Weedon, 1989). Formn.: Formation; Lst: limestone; Non-Lst: calcareous shale and marl (non-limestone). For comparison with the other spectral methods, the results for the Belemnite Marls are also shown even though the % calcium carbonate data do not correspond to square waves. Note that the power values were smoothed using three applications of the discrete Hanning window in order to produce estimates with eight degrees of freedom.

than two lithologies and hence more than two code values are needed, the average Walsh spectrum is affected by the choice of code values (Van Echelpoel, 1994). In this case, either the lithologies should be grouped on geological grounds into two categories or traditional Fourier spectra could be used instead (e.g. Weedon and Read, 1995).

3.5   The statistical significance of spectral peaks

Due to the finite length of observational time series and the presence of spectral side-lobes, the theoretically very smooth, continuous parts of power spectra (i.e. white or red noise components) are approximated by the spectral estimates. This means that the spectral values fluctuate around the smooth continuum forming peaks and troughs (e.g. Fig. 3.10). In the type of mixed spectra encountered in cyclostratigraphy, there is a need to distinguish spectral peaks related to the erratic fluctuations of the spectral background from peaks due to environmentally or geologically significant regular cyclicity.

For spectra generated using the direct method, the multi-taper method or the Blackman–Tukey method, the scale of the erratic fluctuations can be estimated assuming that the uncertainties of the spectral estimates follow a χ^2 (or chi-squared) distribution (Priestley, 1981; Percival and Walden, 1993). All that is needed to determine the spectral confidence intervals is the degrees of freedom of the estimates divided by the χ^2 value appropriate to the confidence interval required (Priestley, 1981). Tables containing χ^2 values are widely available (e.g. Table J in Williams, 1984). For example, for estimates with eight degrees of freedom and the 90% confidence interval, the upper limit (using $P = 0.95$) is $8.0/2.733 = 2.927$ and the lower limit (using $P = 0.05$) is $8.0/15.507 = 0.5159$. Thus, in this example, the uncertainty of the spectral estimates would range from about three times to about half of each power value.

If the range of uncertainty of a particular spectral peak value does not overlap with the continuum spectrum, that peak can be considered to be statistically distinguishable from the background. For red noise, a plot of power versus frequency contains large fluctuations around the spectral background where the average power is high (at low frequencies) and small fluctuations where the average power is low. However, by plotting log power versus frequency these fluctuations appear approximately constant in amplitude across the whole spectrum (i.e. the variance has been stabilized, Fig. 3.26). On log power plots, the uncertainty in the spectral estimates can be indicated using a vertical bar indicating the confidence interval applicable to all the spectral estimates (Fig. 3.26). If the data were plotted as power versus frequency, separate confidence interval bars would be required for every estimate – making the plot untidy.

Note that the 90% confidence interval, for example, indicates that there is a 10% chance that the true position of the spectral estimate lies outside the range indicated by the vertical bar. On this type of plot, to assess whether a spectral peak is distinct from the spectral background one imagines placing the confidence interval on the spectral peak value and judging whether the vertical bar intersects the spectral background (Fig. 3.26). In this approach the investigator does not indicate the spectral background, but for the rather small spectral peaks encountered in cyclostratigraphy this makes it very hard to judge the statistical significance of the spectral peaks. In other words, by avoiding an indication of the implied spectral background it is often hard to assess whether a particular spectral peak is really statistically significant.

The multi-taper method is often used with an F-test for periodic components that have constant phase (e.g. see the algorithms of Lees and Park, 1995). Unfortunately,

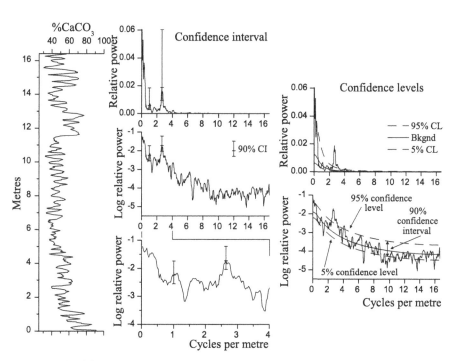

Fig. 3.26 Spectral peaks cannot be used to argue that regular cyclicity is present in a time series until it has been demonstrated that the peaks are statistically distinct from the background spectral noise. This involves calculating confidence intervals based on the chi-squared distribution using the degrees of freedom of the spectral estimates. Two procedures are used to illustrate the results: plotting confidence intervals or plotting confidence levels.

When using confidence intervals, plotting power versus frequency is usually unsatisfactory since the size of the confidence intervals is related to the power of each spectral estimate. Instead confidence intervals are usually shown as a single vertical bar on a log power versus frequency plot; that way the bar can be applied to individual log power spectral estimates. In the figure the bar has been illustrated for two spectral peak estimates at around 1.0 and 2.8 cycles per metre.

An alternative procedure is to plot confidence levels. Firstly the nature of the spectral background needs to be inferred (see Figs. 3.29 to 3.31) and located. Given the background location the confidence interval can be applied. Then all that is needed is to identify whether spectral peaks lie above the upper limit of the background confidence interval. Since only the upper limit of the confidence interval is required, it is common practice to simply illustrate it as a confidence level. Confidence levels are just one-sided limits of the chi-squared distribution, so the upper limit of the 90% confidence interval is equivalent to the 95% confidence level. Hence, there is a 95%, or 19 in 20 chance, that spectral peaks above this level are distinct from the spectral background.

if there may be undetected stratigraphic gaps in the section being investigated (Section 5.3.3) there may be periodic components present which do not have a constant phase. In some situations it may be useful to assume that the spectral background is flat or corresponds to white noise. Various procedures are available for estimating the confidence levels associated with white noise models (Fisher, 1929;

Nowroozi, 1967; Pardo-Igúzquiza and Rodríguez-Tovar, 2000). However, as demonstrated in Section 6.2, virtually all cyclostratigraphic time series have a red noise background.

There are a host of factors that make locating a red noise spectral background difficult from first principles (Chapter 5). These include the type of processes affecting the original environmental system, the effects of post-depositional processes such as bioturbation and diagenesis, the effects of data collection and processing (e.g. sample homogenization and interpolation) and whether any regular cyclicity is present in the record (which needs to be known in advance of the tests for significant spectral peaks). Instead the best that can be hoped for is to approximate the spectral background in a reasonably objective manner.

If the spectral background can be approximated, a more satisfactory display method than using confidence intervals for spectral estimates is to illustrate confidence levels applied to the spectral background (Fig. 3.26, Schwarzacher, 1975). Since only spectral peaks, not troughs, are of interest, only the upper limits of confidence intervals need be plotted. Confidence levels correspond to the (one-sided) limits of confidence intervals. Thus the upper limit of the 90% confidence interval is equivalent to the 95% confidence level. Spectral peaks emerging above confidence levels attached to the estimated spectral background can be readily identified (Fig. 3.26). An advantage of using this type of plot is that the power can be shown using a linear scale with the confidence levels illustrated as diverging curves.

The size of confidence intervals or the relative positions of confidence levels depend on the confidence level used (e.g. 95%, 99%, etc., Fig. 3.27) and the degrees of

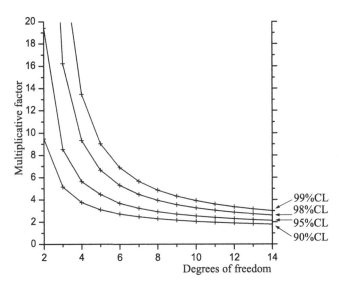

Fig. 3.27 The multiplicative factor used to locate confidence levels depends on the degrees of freedom of the spectral estimates and the probability level (Bloomfield, 1976).

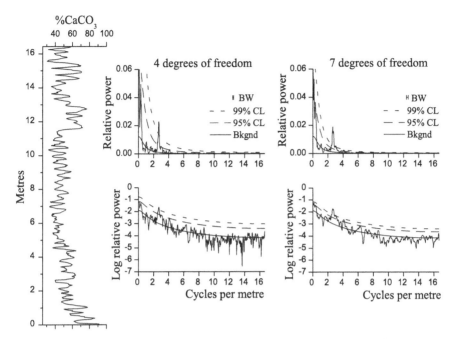

Fig. 3.28 The degrees of freedom of spectral estimates affects the size of confidence intervals and hence confidence levels. The greater the degrees of freedom the smoother the spectrum, the lower the variance of the estimates and the narrower the confidence intervals.

freedom of the spectral estimates (Fig. 3.28). The greater the degrees of freedom, the less the spectral estimates are spread erratically around the spectral background. The degrees of freedom depend on the type of spectral estimation method, whether the data were tapered prior to spectral estimation (i.e. excluding the multi-taper and maximum entropy method) and the degree of zero-padding (Bloomfield, 1976).

There are several approaches available for estimating the position of spectral backgrounds. Most involve using log power versus frequency since this stabilizes the variance of the estimates around the spectral background. A simple method is to estimate the background spectrum by estimating the spectrum using a large number of degrees of freedom so that no peaks and troughs appear (Fig. 3.29). This method is inappropriate for the direct-method and BTM spectra since the largest spectral values are underestimated due to the bias. Unfortunately, the bias increases with larger degrees of freedom. As a result the χ^2 distribution is no longer appropriate for the estimates since this distribution assumes a small bias (Priestley, 1981). The effect of the bias is to cause very smooth spectra to underestimate the position of the continuum at low frequencies, so causing unrealistically high significance to be associated with any low-frequency spectral peaks. A different problem is that the degree of smoothing used to approximate the spectral background is chosen subjectively.

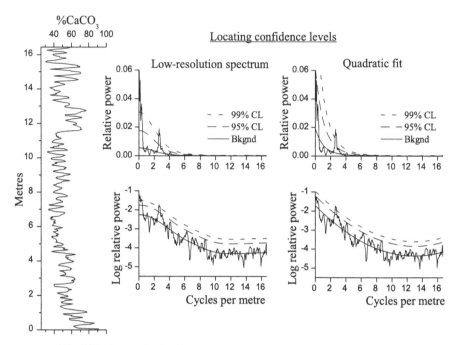

Fig. 3.29 Two methods of locating spectral backgrounds are illustrated: (a) using a low-resolution spectrum (i.e. with very high degrees of freedom) and (b) using a quadratic fit to the log power versus frequency values.

A quadratic (second-order polynomial) fit of the log power versus frequency data is a versatile method for approximating the spectral background since it can be used even if the data possess a very strong autoregressive or moving-average component (Fig. 3.29). Moving-average components are introduced by interpolation (Schulz and Stattegger, 1997). However, if there are large spectral peaks present (i.e. peaks distinct from the background), these can bias the position of the quadratic fit. In other words, the large spectral peaks raise the fitted line, effectively causing the confidence levels to be overestimated and consequently the significance of any spectral peaks to be underestimated.

Mann and Lees (1996) argued that a large proportion of the spectra of climatic and climatic-proxy data have backgrounds corresponding to a first-order autoregressive process. The spectral background can be approximated by using an autoregressive model (see Mann and Lees, 1996 for the equation of an AR(1) spectrum) incorporating the lag-1 autocorrelation (ρ_1) of the data. This has been termed raw AR(1) modelling in Fig. 3.30. However, any regular components in the time series have the effect of biasing the autocorrelation so that ρ_1 is overestimated (Mann and Lees, 1996). To avoid this problem, the median value of the spectral estimates is determined in a short moving window, which is narrower than a quarter of the spectral width and wider than the bandwidth. This 'median-smoothed' spectrum is then used for 'robust fitting' of a first-order autoregressive model by least squares (Mann and

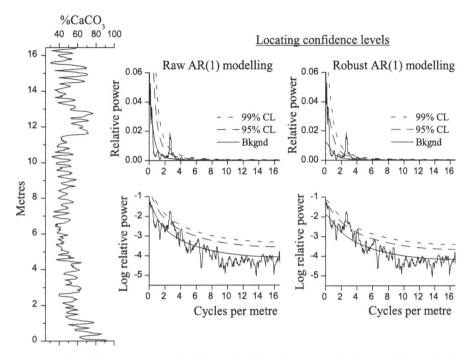

Fig. 3.30 Comparison of confidence levels based on spectral backgrounds located using a raw AR(1) model and located using a robust AR(1) model. The raw AR(1) model is determined by using the lag-1 autocorrelation coefficient of the time series (see equation in Mann and Lees, 1996). In the robust AR(1) model initially the spectrum is replaced by the median values of the spectrum in a moving spectral window. An AR(1) model is then located using a least-squares fit to the median-smoothed spectrum (Mann and Lees, 1996).

Lees, 1996). To further minimize the effect of spectral peaks on the location of the spectral background, the authors also incorporated 're-shaping' of the spectrum (specific to the multi-taper method) into their procedure. In Fig. 3.30 for the robust AR(1) modelling example, the spectrum uses the multi-taper method, but without re-shaping. Robust AR(1) modelling of the spectral background is currently gaining popularity in the cyclostratigraphic community. However, it should be borne in mind that not all spectra can be treated in this manner. Specifically, where there is a strong moving-average component to the data (the log power versus frequency is convex-up rather than concave-up), a robust AR(1) model cannot be fitted to the whole spectrum.

Some authors advocate using log-log plots as most suitable for distinguishing spectral peaks from the spectral background (e.g. Wunsch, 2000a). Thus another way to approximate the spectrum if an AR(1) background is present is to use the log power versus log frequency plot and fit a straight line to the estimates (Fig. 3.31). This corresponds to treating the background as forming a power law (Section 3.3.1). In order to minimize the bias due to any large spectral peaks, a robust linear-fit method

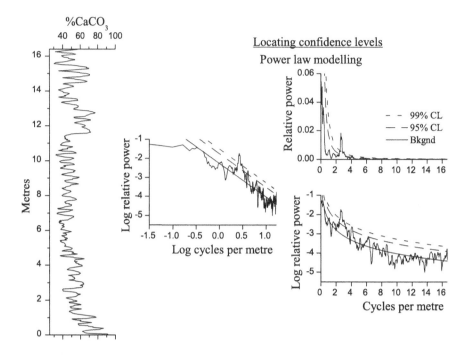

Fig. 3.31 Locating the spectral background using a power law involves initially plotting the spectrum using log-log axes. A robust linear fit is used to determine the position of the spectral background and then confidence intervals and confidence levels can be located in the normal manner.

can be used. For example, Press *et al.* (1992) provide an algorithm that starts with a least-squares fit and then treats data 'too far' from the initially fitted line as outliers (in this case the outliers are the large spectral estimates). Again this method cannot be used when there is a strong moving-average component.

In Fig. 3.32 the approximate confidence levels of spectra for various noise models are illustrated. This is partly designed to show that different methods for approximating the spectral background are needed for different forms of background. But equally important is the well-known observation that even data lacking regular cyclicity generate spectral peaks that are statistically distinct from the spectral background. This emphasizes that in any situation the significance of a spectral peak should be treated cautiously – there is always a finite chance that a particular spectral peak has arisen by chance.

Some statistically significant spectral peaks can arise because of the non-sinusoidal shape of the time series oscillations. Furthermore, there are other issues that need to be addressed before a statistically significant spectral peak can be interpreted as indicative of some form of environmentally or geologically significant regular cyclicity. Thus discussion of the geological or environmental significance of spectral peaks is postponed until Chapter 5 (Section 5.4.2).

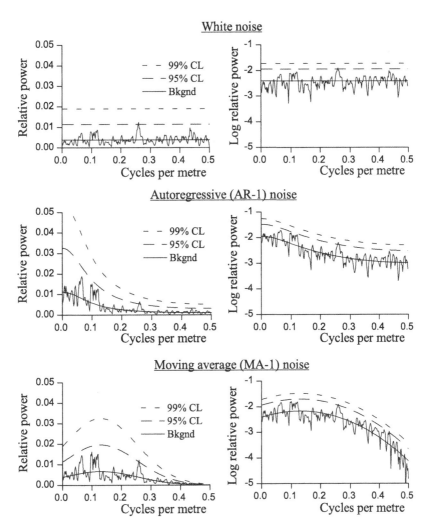

Fig. 3.32 Spectra with different spectral background shapes require different methods for locating the background spectrum that is required for finding confidence levels. The data for the noise models used here are illustrated in Fig. 3.10. First-order autoregressive noise can be modelled using the robust AR fitting procedure of Mann and Lees (1996). This allows the estimation of the average power and ρ_1. White noise is modelled by fixing $\rho_1 = 0.0$. Moving-average noise has a convex-up log power versus frequency spectrum and cannot be modelled using the robust AR-fitting method. Instead a quadratic fit can be used to approximate the spectral background.

Notice that even for data generated without any regular cyclic components, a few spectral peaks can appear significant. This emphasizes the statistical uncertainty of identifying individual spectral peaks as significantly distinct from the spectral background.

3.6   Chapter overview

■ Time series require pre-processing before spectral analysis. The usual minimum requirement is that the series is linearly detrended before processing.

■ Meaningful spectral analysis requires the analysis of time series that have, at least approximately, stationary mean and variance.

■ The discrete Fourier transform simply involves multiplying the time series by a cosine wave having unit amplitude and a particular frequency, and then summing the resulting values. This is repeated for a unit-amplitude sine wave of the same frequency. The square of the result for the cosine wave, plus the square of the result for the sine wave, yields the periodogram power for that particular frequency.

■ The periodogram is not a consistent estimate of the power spectrum – some form of smoothing is required.

■ The direct method of spectral estimation involves applying a weighted moving average to the periodogram values. Alternatively, the spectrum can be estimated using the average of periodograms from non-overlapping subsets (the Bartlett method) or overlapping subsets of the time series (the Welch method). In the multi-taper method the time series is pre-processed using special data tapers. After each taper has been applied the periodogram is generated and then the spectrum is estimated using the mean of the multiple periodograms.

■ In the Blackman–Tukey method the spectrum is estimated using the Fourier transform of the truncated autocovariance sequence. The maximum entropy method involves use of the autocovariance sequence out to the truncation point, but assuming maximum disorder in the time series for autocovariance estimation at higher lags.

■ The Walsh method of spectral estimation is based on analysis using the square-wave Walsh functions rather than sine and cosine functions. The Walsh spectrum depends on the phase of the time series, unless the spectrum is estimated using repeated data shifting to yield the average Walsh spectrum.

■ Confidence levels are often more useful than confidence intervals when judging whether a spectral peak is statistically distinct from the spectral background. Their use requires preliminary estimation of the spectral background using an appropriate noise model.

Chapter 4

Additional methods of time-series analysis

4.1   Introduction

This chapter is principally concerned with techniques for analysing regular cyclicity previously detected via spectral analysis (Chapter 3). However, wavelet analysis, phase portraits and singular spectrum analysis discussed here can be used meaningfully prior to generating a power spectrum. In this, and subsequent chapters, the power spectra were obtained using the multi-taper method with confidence levels from robust AR(1) modelling following the procedure of Mann and Lees (1996). As for Chapters 2 and 3, sources of computer algorithms are listed in the Appendix.

4.2   Evolutionary spectra

It is often difficult to be sure whether a time series is stationary in terms of the variance (Section 3.2.2). Power spectra indicate the average variance density versus frequency for the whole time series and this may give a misleading impression of the data. Clearly, if regular cyclicity of a particular wavelength is sought, it is important to be sure that it has a large amplitude throughout the data set (in this case meaning an amplitude that is significantly different from the background levels). A simple test can be performed, by generating spectra for the two halves of the time series, and then checking for the persistence of a spectral peak at the expected frequency. A disadvantage of this method is that the bandwidth of the two subspectra is double that of the spectrum for the whole record.

As an example, the spectra of three segments of the whole of the Belemnite Marls carbonate record, excluding the aliased data identified in Fig. 2.9, are compared in

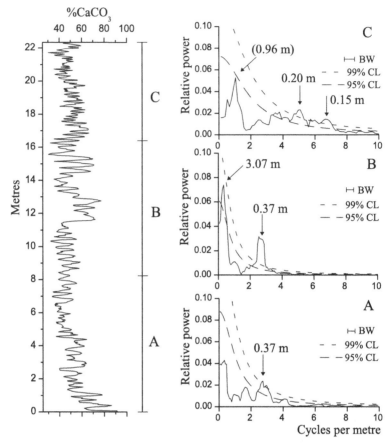

Fig. 4.1 The spectra for successive intervals of the complete Belemnite Marls carbonate data (excluding the aliased data shown in Fig. 2.9). The results indicate shorter wavelength couplet cycles above the 16.41 m level. The cluster of significant peaks in the high-frequency part of the spectrum for segment C occurs because of the decrease in couplet wavelength within the segment. The lower two-thirds of the data are essentially stationary since the same 0.37 m spectral peak is detected in segments A and B. The decrease in couplet wavelengths in the upper part of the Belemnite Marls has been attributed to decreasing accumulation rates (Weedon and Jenkyns, 1999).

Fig. 4.1. It is clear that the substantial decrease in wavelength of the bedding couplets above the 16.41 m level is detectable in the spectrum of the uppermost data segment since the couplet peak has moved to higher frequencies. Conversely, the lower two-thirds of the data are stationary since the same spectral peak, related to cycles with a wavelength of 0.37 m, occurs in each spectrum. When time series are long in comparison to the scale of the regular cyclicity of interest (e.g. more than 20 repetitions of the cycles), then it is feasible to generate many spectra from many subsections and still have a sufficiently narrow bandwidth for cycle detection. This is known as **evolutionary spectral analysis**, **sliding-window spectral analysis** or **windowed Fourier analysis**. In geophysics evolutionary spectra are called **spectrograms** (and in zoology records of animal calls and songs are analysed using **sonograms**). Early examples of this method, in the context of cyclostratigraphy, were applied to deep-sea oxygen isotope time series and wireline-log data (Pisias and Moore, 1981; Molinie *et al.*, 1990).

Evolutionary spectral analysis can be characterized by the use of either consecutive or overlapping data subsections. When the subsections do not overlap, there can be an independent evaluation of the changing nature, or otherwise, of the complete time series. Most commonly, the spectra are based on overlapping data segments on the assumption that there are gradual changes in the frequency distribution of the power through the time series. Abrupt changes in power distribution cannot be resolved with overlapping subsections; for step-like changes in cycle frequency, the overlapping segments produce a low stratigraphic or temporal resolution. As a result, instead of one spectral peak showing a jump in frequency, the evolutionary spectrum will contain two peaks at the level of the step. Sometimes the spectra from overlapping segments are illustrated as a series of offset spectral plots (e.g. Berger, 1984; Yang and Baumfalk, 1994). A more satisfactory method is to plot frequency against the stratigraphic position of the centre of each segment, and then contour the power values, perhaps adding shading or colour (Fig. 4.2, e.g. Melnyk *et al.*, 1994; Schaaf and Thurow, 1997; Meyers *et al.*, 2001).

Since evolutionary spectra are usually based on relatively short data segments, it makes sense to use the highest-resolution spectral method available (Chapter 3). After choosing a spectral estimation technique, three other parameters need to be considered. The degree of spectral smoothing (degrees of freedom) influences the amplitude of the spectral noise as well as the frequency resolution. The length of the data segments influences the frequency resolution and, as already mentioned, the degree of segment overlap affects the stratigraphic resolution of the plot. Thus there is an inevitable trade-off between high frequency resolution and high temporal or stratigraphic resolution, a form of uncertainty principle (Mallat, 1998).

In the case of the Belemnite Marls shown here a contoured multi-taper method evolutionary spectrum is illustrated (Fig. 4.2). For the lower part of the evolutionary spectrum there are two spectral peaks which form ridges at around 0.35 and 2.5 cycles per metre. The lower frequency ridge is related to the metre-scale bundle cycles (wavelength around 3 m), while the higher frequency ridge relates to the bedding couplets (wavelength around 0.4 m). Additional peaks that come and go are regarded here as noise. Above the 16.41 m level the two ridges shift towards higher frequencies. A step-like shift in cycle frequency/wavelength would have created a zone of multiple peaks around the 16.41 m level. In such a circumstance the cycles in the upper part of the Belemnite Marls might be considered to be unrelated (i.e. of different environmental significance) to those below. Instead, in this case the simultaneous shift of both power ridges towards higher frequencies at the same stratigraphic level is more plausibly interpreted as indicating gradually decreasing accumulation rates (cf. Herbert, 1994).

Evolutionary spectra are also used for studying data with a time, rather than a thickness, scale. In this case, a change in the frequency of a spectral peak or ridge would

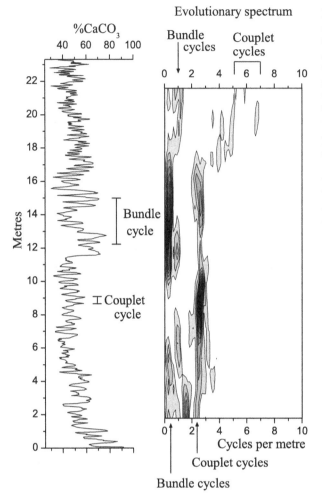

Fig. 4.2 A contoured evolutionary spectrum for the Belemnite Marls carbonate data. The plot was constructed using time series segments 3.45 m long with a step size of 0.90 m. The successive multi-taper method spectra were plotted, relative to the complete time series, stratigraphically half way through each overlapping time series segment. The contouring is in 0.02 units of relative power.

indicate either a problem with the age-depth model used or a change in the frequency of environmental cyclicity (such as the shift from 41 ka to 100 ka cycles in the early Pleistocene, Ruddiman *et al.*, 1989). On evolutionary spectra, the changes in the height of the power ridges can be used to monitor changes in the amplitude of a particular regular cycle. However, filtering and complex demodulation are more appropriate for monitoring amplitude variations when the frequency of the cycle is more or less constant, since the latter techniques yield higher temporal or stratigraphic resolution.

When there is a very strong red noise component in the individual spectra, evolutionary spectra become dominated by low-frequency power. This problem can be overcome by: (a) pre-whitening the data (Section 3.2.4); (b) high-pass filtering the data (Section 4.3, e.g. Melnyk *et al.*, 1994); or (c) finding the spectral background for each individual component spectrum and then contouring in terms of spectral peak significance rather than power or relative power (e.g. Dunbar *et al.*, 1994; Rittenour *et al.*, 2000).

4.3   Filtering

Filtering involves manipulating a time series so that there is a change in the spectral characteristics of the data. Broadly, filtering is divided into **frequency-selective filtering**, **threshold filtering** and **Wiener filtering** (Fig. 4.3, Ifeachor and Jervis, 1993). In each case a different procedure is used to isolate the signal from the 'noise' in the time series. Frequency-selective filtering involves removal of frequency components that are not wanted (Fig. 4.4). Threshold filtering involves removing all the information related to variations up to a certain threshold power level (or amplitude level). However, when there is a substantial spectral background upon which spectral peaks are superimposed, Wiener filtering is appropriate. Wiener filtering can be used to model the spectral background and then remove only a component of the spectral power from a wide range of frequencies (Priestley, 1981; Press *et al.*, 1992; Ifeachor and Jervis, 1993). This procedure could be employed to remove the continuum component of a

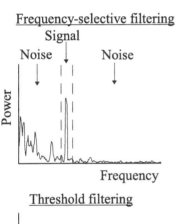

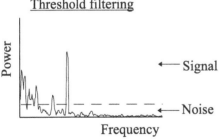

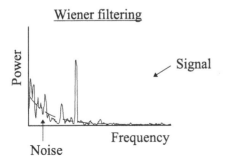

Fig. 4.3 Types of filtering. To date cyclostratigraphic studies have employed frequency-selective filtering almost exclusively.

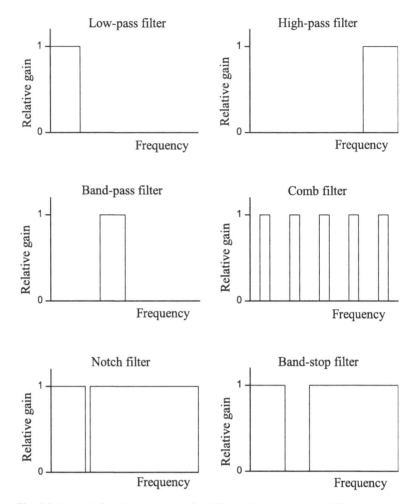

Fig. 4.4 The gain function measures, for different frequencies, the difference in amplitude of the filter output compared to the original time series. This figure shows ideal gain plots for frequency-selective filters.

mixed spectrum, leaving the statistically distinct spectral peaks. However, Wiener filtering has yet to be used extensively in cyclostratigraphic studies. Instead the dominant method of filtering to date has been frequency-selective filtering.

There are a variety of ways in which frequency selection can be achieved (Fig. 4.4). The most commonly used methods in cyclostratigraphy are **low-pass**, **high-pass**, and **band-pass filtering**. These techniques result in the retention, in the spectra of the filter output, of low- or high-frequency power or a narrow band of power respectively. Frequency-selective filters treat the '**pass-band**' information as signal and the 'rejected' information in the '**stop-band**' as noise to be removed or 'filtered out' (Ifeachor and Jervis, 1993). Thus low-pass filters are used to isolate the low-frequency components of a time series. Band-pass filters are useful for isolating individual regular cyclic components, or a single delimited range of frequencies. For comparison, in electronic sound systems the treble control acts as a high-pass filter and the bass control as a low-pass filter (needed at low volumes to compensate for the non-uniform frequency response of the ear and brain, Taylor, 1976). Sometimes the width of the filter pass-band is called the **filter bandwidth**.

For frequency-selective filtering, there are a number of important considerations. Firstly, any time series, even random numbers, can be band-pass filtered to restrict the range of frequencies present. Thus, it is essential that the choice of frequencies is justified. This is a way of saying that band-pass filtering should only be used to study time series where there are statistically significant spectral peaks. Otherwise, the filtering would merely isolate a restricted range of frequencies from the continuum of noise, creating a misleading impression of regular cyclicity. Note that even if a spectral peak, used to justify filtering, is significant, a large spectral background within the range of the pass-band will lead to a substantial noise component in the output. In this situation Wiener filtering is more appropriate.

Ideally, frequency-selective filters are designed by defining the edges of the pass-band. However, in Fourier techniques it is impossible to achieve an abrupt, step-like change in amplitude as a function of frequency from the pass-band to the stop-band. Instead, a **transition-band** is always present and its width also needs to be considered during filter design (Fig. 4.5). The **gain function** defines how the amplitudes of the time series components change during filtering. Note that gain refers to change in amplitude as a function of frequency, not the change in power. Filtering can cause a change in phase of the time series without a significant change in shape (e.g. a shift in time or stratigraphic position, Section 5.3.2). Depending on the type of filtering the change in phase may vary according to frequency. The equation that defines the changes in amplitude and phase is called the **transfer function** of the filter (Priestley, 1981).

The design of filters is partly concerned with ensuring that the gain is close to one throughout the pass-band and close to zero in the stop-band. In other words, the gain function would, ideally, have a flat top and base. Unfortunately, most types of filtering procedures generate small ripples in the gain function which also need to be controlled (caused by the Gibbs' phenomenon, Ifeachor and Jervis, 1993). Thus it is wise to check the shape of the gain function after filtering. In Walsh analysis (Section 3.4.6), the discontinuities of the Walsh functions mean that there is no difficulty in producing filters that have gain functions with sharp transitions from the pass-band to the stop-band and there are no ripples in the gain function (Beauchamp, 1984; Weedon, 1989).

A particularly simple type of low-pass filter, available for uniformly spaced records, is exemplified by a moving average. For example, consider the familiar three-point Hanning weights 0.25, 0.50, 0.25. Applied to a unit impulse seven points long (i.e. 0, 0, 0, 1, 0, 0, 0), this three-point moving average produces an output: 0, 0, 0.25, 0.5, 0.25, 0, 0. The moving average just consists of applying the weights to the original data using, in this case, three values at a time, and moving one point along the record at each step. Applied twice, the output would be 0, 0.0625, 0.25, 0.375, 0.25, 0.0625, 0. This method of filtering in the time domain is called **convolution**. The moving-average weights amount to **filter coefficients**. When applied to the unit impulse, as done in this

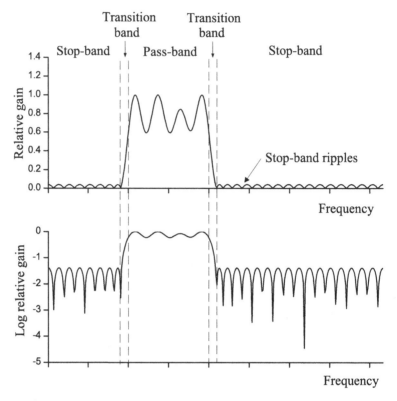

Fig. 4.5 The terminology applied to the gain function of a band-pass filter. Note that in real frequency-selective filters there is always a transition band separating the pass- and stop-bands (cf. Fig. 4.3).

example, the output is called the **impulse response**. This example shows that, aside from the zeroes, the output values when filtering a unit impulse (the impulse response) are identical to the filter coefficients. Applying the three Hanning weights twice is the same as using a series of five weights (the non-zero output values in the example) once. Band-pass filters can be constructed by subtracting the output of one low pass filter from another. Note that pre-whitening (Section 3.2.4) is a form of high-pass filtering.

In practice the filter coefficients are more numerous than in this example and the absolute values need to be determined to a high degree of precision. Nevertheless, inspection of the filter coefficients/impulse response gives a good idea of the frequencies that will be left relatively unaltered by the filter. In fact the amplitude spectrum of the impulse response, suitably zero-padded, is the gain function. The impulse response can also be obtained by using an inverse Fourier transform of the transfer function. Figures 4.6 and 4.7 illustrate low-pass and band-pass filtering of the Belemnite Marls data generated using the algorithms of McClellan *et al.* (1973, explained in detail by Ifeachor and Jervis, 1993). Note the similarity in the scale of the oscillations in the respective impulse response and output.

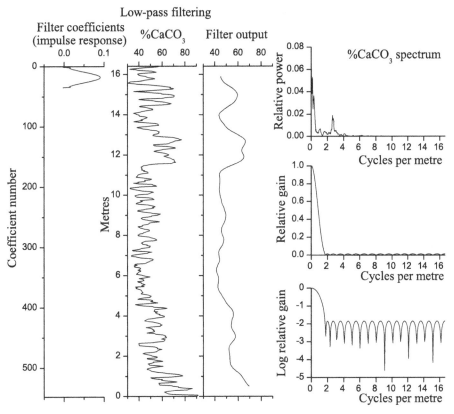

Fig. 4.6 Low-pass filtering applied to the lower two-thirds of the Belemnite Marls carbonate data. The filter coefficients (i.e. the impulse response) are convolved with the data to produce the filter output. Note that in this case the filter coefficients essentially consist of a single low-frequency oscillation. The gain plot shows that an ideal low-pass filter has not been attained since the gain function has a fairly broad transition band. However, there is satisfactory suppression of the high-frequency oscillations in the data. This is demonstrated by the log relative gain plot where the maximum gain in the stop-band is just 1% of the original amplitude (i.e. the log gain is −2.0).

As mentioned, the type of filter implementation in the time domain just described is termed convolution. This can be thought of, from a signal processing perspective, as taking the incoming time series and modifying it as it passes through the filter. If the input signal is turned off, the output soon dies away to nothing. In such filtering the impulse response is much shorter, in terms of number of coefficients, than the length of the data. Hence this represents a **finite impulse response** or **FIR filter**. Alternatively, the time series can be filtered while the data are received, but the filtered output can include a certain component of the output generated earlier. In this case if the input is turned off, the output might carry on indefinitely since the filter can act on the earlier output – in the absence of any new input. Thus this type of filter is described as an **infinite impulse response** or **IIR filter** (Ifeachor and Jervis, 1993). For cyclostratigraphic data, either FIR or IIR filters can be used.

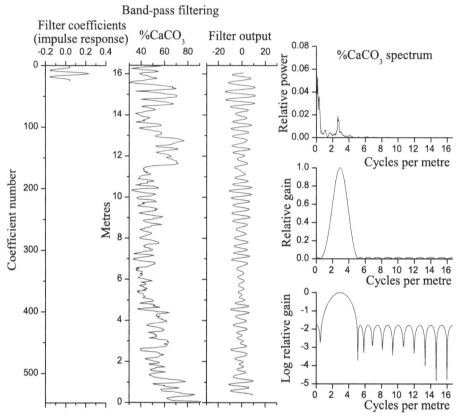

Fig. 4.7 Band-pass filtering applied to the lower two-thirds of the Belemnite Marls data, in this case used to extract the regular couplet cycle. Note that filter design should always be based on prior spectral analysis – in this case it is known that there is a significant spectral peak centred at 2.7 cycles per metre. The filter coefficients of band-pass filters involve oscillations comparable in wavelength to the regular cyclicity.

One aspect of filtering that should always be checked is the phase of the output. Some filters change the phase so that, for example, where maxima are expected in the output, minima or an intermediate phase of the oscillations is encountered instead. The ease with which the **phase shift**, or **phase delay**, can be corrected depends on the type of filter. FIR filters that have a symmetrical impulse response produce linear phase shifts. This means that the same shift occurs at all frequencies so it is easily corrected (Ifeachor and Jervis, 1993). IIR filters produce non-linear phase shifts. Some authors plot band-pass filter output directly on top of the original time series, so that the gain and phase can be assessed visually (e.g. Ruddiman *et al.*, 1989). Certainly, the output should always be shown alongside, if not on top of, the original data, preferably using the same amplitude scale for both plots. Note that band-pass filters that are too narrow tend to produce output with spuriously slowly varying amplitude modulations. In other words, overly narrow band-pass filters fail to capture the true amplitude modulation of the target frequency band – hence the importance of comparing the output with the original data.

The practical design of a filter, once the choice of FIR or IIR has been made, can also be rather complex (Otnes and Enochson, 1978; Ifeachor and Jervis, 1993; Emery and Thomson, 1997). One method for FIR filter design involves defining the transfer function that is desired and then using an inverse Fourier transform to specify the impulse response or coefficients required. FIR and IIR filters used in the time domain are known as **causal filters**, since they can be implemented by real physical systems in 'real time'. In other words, the filters act on information from the present and past only. An alternative approach to filtering is to use an **acausal filter** where all the data are manipulated at once.

For example, an acausal filter could involve multiplying the Fourier transform of the data in the frequency domain by a set of weights that reach unity in the pass-band and zero in the stop-bands, and then inverse Fourier transforming the results (Press *et al.*, 1992). (Convolution of filter weights and a time series is equivalent to simple multiplication of the Fourier-transformed filter weights (i.e. the transform function) and the time series in the frequency domain.) Using this procedure, the sharp edges of the pass-band in the frequency domain can lead to spurious oscillations appearing in the output. Effectively, the transfer function (frequency domain weights) would need to be repeatedly modified until a satisfactory output was obtained. Note that, as previously mentioned, the Walsh implementation of this form of acausal filtering does not suffer from these problems (Beauchamp, 1984). For comparison, the advantage of the time-domain filtering is that it allows the immediate specification of the final characteristics of the transfer function (Ifeachor and Jervis, 1993). In non-linear time-series analysis, singular spectrum analysis (Section 4.8) can be used to implement a form of acausal FIR filter (Kantz and Schreiber, 1997). Both causal and acausal filtering have trouble with the abrupt ends of real data sets. For example, during convolution an FIR impulse response that is an odd integer or N points long cannot yield output less than $(N-1)/2$ points from the ends of the data (Ifeachor and Jervis, 1993). Consequently the ends of the data cannot normally be filtered.

4.4   Complex demodulation

As just explained, band-pass filtering can be used to extract regular cycles from time series. This allows variations in amplitude (amplitude modulations) to be visualized. However, sometimes it is useful to be able to determine the continuous variations in amplitude throughout the time series (i.e. at peaks, troughs and intermediate positions). This can be achieved by using an algorithm based on complex numbers, hence the term **complex demodulation** (Bloomfield, 1976). The investigator nominates the frequency at which demodulation is required and applies the algorithm to the band-pass-filtered record. The output is initially very noisy as it includes information at all frequencies. However, the method is designed around the assumption that the change in amplitude occurs gradually. Thus the output merely needs to be low-pass filtered so that no oscillations are left at the scale of the cycles of interest (Fig. 4.8).

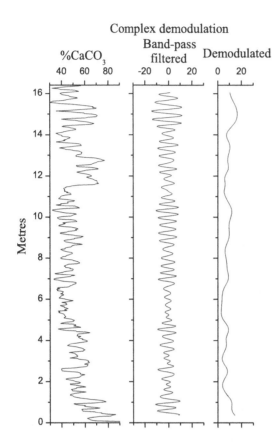

Fig. 4.8 Complex demodulation allows the estimation of the continuous amplitude variations (amplitude modulations) of a particular frequency. Here the data and results for the Belemnite Marls are shown using horizontal axes with the same scaling to allow direct comparison between the filtered output, the demodulated output and the original time series.

As for all filtering, the success of this low-pass filtering should be checked via the spectrum of the output and the gain function. The result is an oscillating output that reflects the variations in amplitude at the frequency component of interest. In AM radio and in some cyclostratigraphic studies, demodulation is achieved by half-wave rectification (e.g. by truncating all the values of the band-pass-filtered data above the mean) followed by low-pass filtering (Olsen, 1977). This cruder method of demodulation is only really appropriate when the amplitude variations of the band-pass-filtered output are symmetrical about the mean (often not true for cyclostratigraphic data).

Taner *et al.* (1979) investigated complex demodulation in the context of seismic data analysis. Pisias and Moore (1981) were perhaps the first to use complex demodulation in a cyclostratigraphic context. As another example, Pisias *et al.* (1990) showed, using complex demodulation, that the response time of the climate to orbital-insolation cycles must have varied during the Pleistocene. Additionally, Shackleton *et al.* (1995c) demonstrated that complex demodulation can be more useful than coherency spectra for checking the success of the orbital tuning of time series from deep-sea sediments (Section 6.9.3). The Bloomfield (1976) complex demodulation algorithm also yields the variations in phase, or **instantaneous phase**, of the frequency component of interest (again low-pass filtering is necessary). It might be thought that a record of demodulated phase could be used to check for phase changes due to otherwise undetected hiatuses in time series based on thickness scales rather than time scales. Unfortunately, the smoothing required precludes detection of abrupt phase shifts. Nevertheless, Rutherford and D'Hondt (2000) proved that instantaneous phase can be useful for studying palaeoclimatic records.

4.5   Cross-spectral analysis

It is often useful to be able to compare time series of different variables measured at the same time or on the same samples. This can allow an improved understanding of the environmental system under investigation. Multiple time series can be compared in pairs using **cross-spectral analysis**. The most important types of cross spectra are **coherency** and **phase spectra**.

4.5.1   Coherency spectra

Coherency is effectively a measure of the similarity of the amplitude variations in two time series determined at particular frequencies. It is analogous to a correlation coefficient, but it is measured at one scale of variability (frequency), rather than all scales together, and it can only range from 0 (zero coherency) to $+1.0$ (perfectly coherent).

Coherency is calculated via the **cross spectrum**. Recall from Chapter 3 that, simplifying, autocorrelation is calculated by multiplying a time series by itself for different lags. The autocovariance is just the autocorrelation multiplied by the variance, and the Blackman–Tukey spectrum can be obtained using the Fourier transform of the lag-windowed autocovariance (Section 3.4.4). Similarly, the cross spectrum can be derived by: (a) multiplying together the two time series at different lags (rather than a time series with itself) to obtain the cross-correlation sequence; (b) applying a lag window to the cross-covariance; and (c) using the Fourier transform. The similarity of the two records can then be established by dividing, for each frequency, the cross spectrum by the square root of the product of the power spectra of the two time series (Bloomfield, 1976; Priestley, 1981). Coherency must be calculated using the power spectra and cross spectrum rather than via the periodograms and cross periodogram (Bloomfield, 1976). Note that some authors distinguish coherency from squared coherency – which is also known as **coherence**. Coherency measures the similarity in amplitude variations, whereas coherence measures the similarity in variance variations (analogous to the difference between amplitude spectra and power spectra, Section 1.3). Coherency indicates linear not non-linear relationships (cf. Hagelberg and Pisias, 1990).

The level at which coherency can be considered significantly different from zero depends on both the variance of the spectral window used (which is affected by zero-padding) and the desired confidence level (Bloomfield, 1976). This level of significant coherency is usually indicated using a dashed horizontal line. It has been common for palaeoceanographers to use the 80% level for minimum coherency significance but, by analogy with correlation coefficients, the 90% or 95% levels are more prudent. The uncertainty in coherency estimates is related to the coherency value (i.e. the uncertainty is smaller when the estimated coherency is large, Bloomfield, 1976). Coherency is sometimes plotted using a hyperbolic arc-tangent scale so that a single uncertainty interval can be illustrated for high and low coherency values. In other words, the hyperbolic arc-tangent scale stabilizes the variance of the coherency (analogous to the use of a log scale for plotting a single confidence interval for power spectra, Fig. 3.26).

To aid in the interpretation of coherency spectra, it is normal for the associated power spectra to be plotted above. This allows the frequencies at which regular cyclicity has been identified using spectral peaks to be examined on the corresponding coherency spectra. In Fig. 4.9, 920 seasonal values of central England temperature are compared to England and Wales precipitation for 1766 to 1996 (data from Hulme and Barrow, 1997). Both power spectra reveal strong annual cycles and very high coherency at the frequency of a year. This indicates that annual variations in rainfall are linked, directly or indirectly, to annual variations in temperature.

Coherency can be significant at frequencies where no significant spectral peaks are present. This indicates similar behaviour of the two time series variables, but that the oscillations are due to continuum noise rather than regular cyclicity. It is also possible for there to be coincident and significant power-spectral peaks, but non-significant coherency at that frequency. A somewhat hypothetical example of this might be where total rainfall during a year is related to the intensity of a monsoon rather than the yearly variation in local temperature. The power spectra of temperature and rainfall could both reveal annual cyclicity, but year-to-year variations in rainfall, dictated by regional monsoonal processes, will not necessarily be coherent with year-to-year variations in the local temperature.

Figure 4.10 shows the coherency spectra for sine and cosine waves plus first-order autoregressive noise where the timing of the imposed amplitude modulation (Section 1.3) is the same. The coherency is close to one since the average amplitude variations are similar. Figure 4.11 shows a comparable situation except that two sine waves are compared and the amplitude modulations occur at different times. In this latter case the coherency is still close to one because the nature of the amplitude varia-tions is the same. The difference in timing of the amplitude modulations is not apparent from the coherency plots. Shackleton *et al.* (1995c) argued that, rather than relying on coherency plots, orbitally tuned records (Section 6.9.3) should be demodulated and compared to the amplitude modulation of the tuning target in order to check the success of the tuning. This is because high coherency on its own does not prove that the tuning was 'correct'. Another problem with using coherency is that when the two measured variables, X and Y, involve common factors, (e.g. $X = A/C$ and $Y = B/C$), the coherency is artificially increased. A similar problem arises in the use of correlation coefficients where spurious correlation arises from common factors in the variables being compared (Williams, 1984).

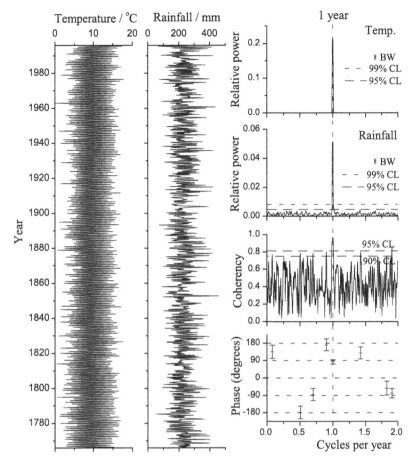

Fig. 4.9 Comparison of two time series is possible using coherency and phase spectra. Here seasonal records of central England's temperature are compared with the precipitation (principally rainfall) of England and Wales from 1766 to 1996 (data listed by Hulme and Barrow, 1997). The power spectra are dominated by the annual cycle. Note that background power levels are so low for the temperature record that the confidence levels are not visible. At the scale of the annual cycle the coherency is very high and the phase difference is $+83.8°$ $±11.0°$. The phase difference means that, within the uncertainty of the phase estimates, over periods of one year changes in temperature lead changes in rainfall by a quarter of a year ($90°$), or one season.

Despite these caveats it is important to recognize that coherency provides a much more sophisticated way to compare two time series than a simple correlation coefficient. This is due to the way that coherency varies according to frequency (different wavelength components are compared) and because the phase of the data is not involved. For example, consider the scatter plots in Fig. 4.12 based on the data in the two previous figures. Visually it is obvious that the two records in Fig. 4.10 are strongly correlated, despite the comparison of sine and cosine waves. This is reflected in the fairly elongate cluster of points in the top part of Fig. 4.12. However, inspection of the bottom part of Fig. 4.12, or the use of the corresponding

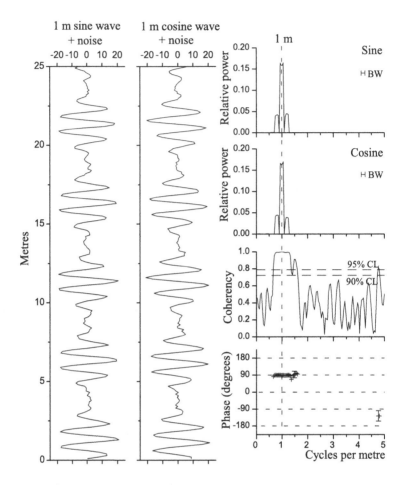

Fig. 4.10 Comparison of amplitude-modulated 1 m sine and cosine waves with the addition of low-amplitude AR(1) noise. Coherency is close to one at the 1 m scale since the amplitude variations are similar. (The AR(1) noise was added to prevent coherency reaching one at all frequencies.) The phase difference is +90° at the 1 m scale since the regular cycles are sine and cosine waves.

correlation coefficient, would not have revealed the similarity in amplitude variations in Fig. 4.11.

As for power spectra, it cannot be assumed that coherency is stationary through a time series (Priestley, 1988). For example, in the bottom third of the Belemnite Marls carbonate and total organic carbon data, the variations are not coherent at the scale of the regular bedding couplets despite the similarity in power spectra (Fig. 4.13). However, in the middle third of these data the variables have similar power spectra and are coherent at the scale of the couplets (Fig. 4.14). The reason for this changing relationship is unclear at present (Weedon and Jenkyns, 1999). Nevertheless, this does demonstrate that at face value a coherency spectrum for the lower two-thirds of the Belemnite Marls data would be rather misleading. This suggests that a plot of evolutionary coherency, analogous to evolutionary power, would be worthwhile.

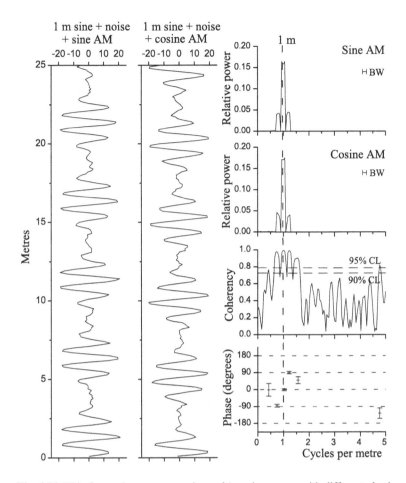

Fig. 4.11 This figure shows a comparison of 1 m sine waves with different phasing for the imposed amplitude modulation (AM). As the amplitude variations are similar at the 1 m scale, even though the timing differs, the coherency is close to unity. As sine waves are being compared the phase difference is zero. In other words, phase measures the relative timing of peaks and troughs at the scale of the frequency in question rather than at the scale of the imposed AM. This shows that the relative timing of amplitude modulations cannot be deduced using coherency or phase spectra – an important consideration in orbital tuning (Shackleton *et al.*, 1995c).

4.5.2   Phase spectra

The average difference in phase between two time series at different frequencies is described by the **phase spectrum** (Bloomfield, 1976; Priestley, 1981). **Phase difference**, commonly abbreviated to **phase**, ranges from $+180°$ to $-180°$ (or in radians, from $+\pi$ to $-\pi$, Fig. 4.15). A zero phase difference indicates oscillations that are 'in phase' whereas for other values the oscillations are 'out of phase'. Positive phase indicates that the first series leads the second, and negative phase that the first series lags the second. Note that $+180°$ and $-180°$ are equivalent (Fig. 4.15) so that phase spectra could be plotted on a cylinder with the cylinder axis parallel to frequency. Since

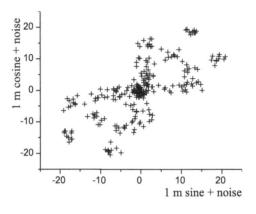

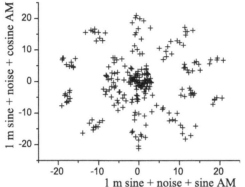

Fig. 4.12 Scatter plots for the data illustrated in Figs. 4.10 (top) and 4.11 (bottom).

the plots are actually two dimensional, trends in phase as a function of frequency can cause sudden jumps from the positive to the negative parts of the phase spectrum. Phase difference is apparently used in human hearing/brain function to help establish the locations of sources of low-frequency sound by comparing the sound received by each ear (Hartmann, 1999).

Phase can be expressed in equivalent time or thickness units using: time delay = cycle period × (phase/360°), or stratigraphic offset = wavelength × (phase/360°). For example, in Fig. 4.9 at the scale of the annual cycle, temperature has a phase of +83.8° ±11.0° relative to precipitation. In this case the phase error (the 95% confidence interval) means that the phase could range from +72° to +94° and is therefore indistinguishable from a phase error of +90°. Hence, over one year, on average temperature changes lead rainfall changes by 0.233 years or 2.8 months, or essentially one season (i.e. within the uncertainty, by a quarter of a year). In Fig. 4.10, the phase difference is +90° for the 1 m cycles because sine and cosine waves were compared (Fig. 1.1). Similarly, at the scale of the 1 m cycles, the phase is 0° in Fig. 4.11 since two sine waves were used. This is because the maxima and minima of the cycles are aligned, or in-phase, regardless of the amplitude.

The uncertainty in phase depends on the coherency and variance of the spectral window (and hence zero-padding); phase errors at low coherency are very large (Fig. 4.16,

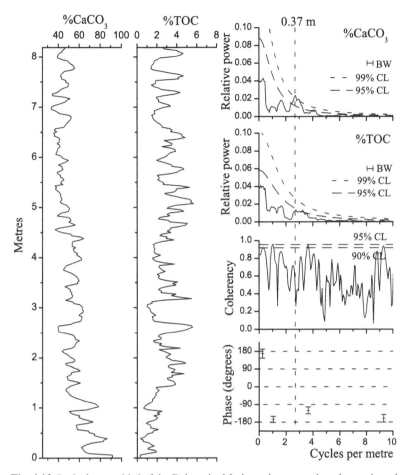

Fig. 4.13 In the lowest third of the Belemnite Marls, carbonate and total organic carbon (TOC) contents are not coherent at the scale of the couplet cycles – despite the presence of significant spectral peaks for both variables at 2.7 cycles per metre.

Bloomfield, 1976). Since the 95% confidence limits for phase estimates exceed $\pm 45°$ (an uncertainty range of 90°) at even moderate coherency, phase values where coherency is medium or low are of limited use. Consequently, it is common for the phase to be plotted only where the coherency exceeds the appropriate coherency confidence level. For the phase spectra in this book, phase estimates are only shown at frequencies where the coherency exceeds the 90% confidence level. On the phase spectra, the phase 95% confidence intervals (vertical bars) are illustrated at the frequency of significant coherency peaks (e.g. Fig. 4.14).

The Belemnite Marls data in Fig. 4.14 show that carbonate and organic carbon contents are inversely related – the phase is indistinguishable from 180°. This relationship remains valid even if allowance is made for the diluting effects of carbonate on the organic carbon (by examining carbonate-free organic-carbon contents,

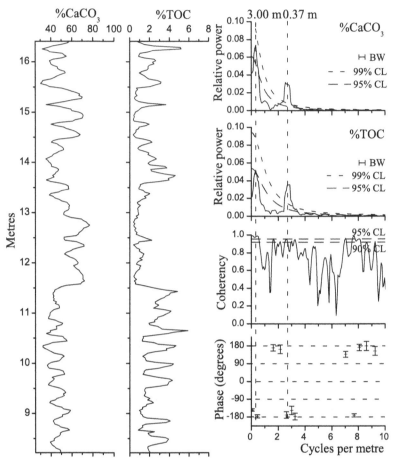

Fig. 4.14 In the middle part of the Belemnite Marls, carbonate and TOC contents are coherent and inversely related (i.e. a phase difference of 180°) at the scale of both the bedding couplets and the bundles. The change in coherency from the lower third of the data shown in Fig. 4.13 indicates a changing relationship between carbonate and organic carbon contents which has yet to be explained (Weedon and Jenkyns, 1999).

Weedon and Jenkyns, 1990). However, for real physical systems where two variables reflect input and output, the maximum possible phase difference is 90° (Pisias *et al.*, 1990). Thus some authors invert one of the data sets (multiply the data by minus one) before cross-spectral analysis so that only phase differences up to ±90° appear (Shackleton, pers. comm. 2001). This is commonly applied to $\delta^{18}O$ measurements because lower values correspond to larger ice volumes in the late Neogene (e.g. Hall *et al.*, 2001). In the case of the Belemnite Marls this would require inverting the scale of one of the variables (e.g. %TOC) so that the phase would be indistinguishable from 0°. Nevertheless, the physical interpretation would be unchanged: low carbonate contents are associated with high organic-carbon contents. A further consideration is the comparison of rates of processes. Remember that, for sinusoidal oscillations,

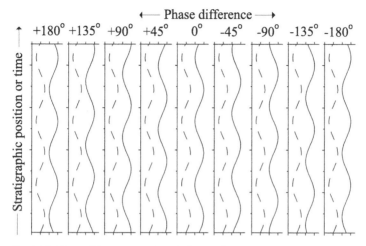

Fig. 4.15 Phase difference or phase indicates the relative timing of two cycles that share a particular frequency. The phase differences indicated are relative to the dashed curves (NB: the oldest data occur at the base of the plots).

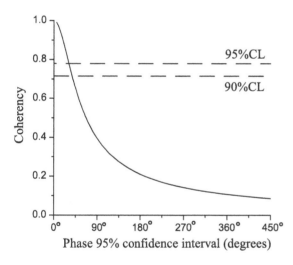

Fig. 4.16 The uncertainty (95% confidence interval) in phase values is very large when coherency is small. As well as coherency, the phase confidence interval depends on the variance of the spectral window (Bloomfield, 1976) so the graph shown represents a particular case. Nevertheless, it is clear that when the coherency is below the 90% confidence level (horizontal dashed lines) the uncertainty is too large for the phase values to be of much use since the confidence interval exceeds ±45°.

the rate of change of a variable has a phase that is 90° ahead of the original variable (e.g. a maximum in the rate of change occurs before, or leads, the maximum in the variable).

The combination of power-, coherency- and phase-spectra can be extremely useful for studying environmental systems when many variables have been observed using the same samples. The power spectra reveal the presence and periods or wavelengths of any regular cyclicity, coherency indicates which variables are likely to be linked environmentally, and the phase spectrum reveals the relative timing of changes in pairs of variables. This allows a detailed investigation of environmental processes and their relationships (e.g. Imbrie *et al.*, 1984; Weedon and Shimmield, 1991). Sometimes to help clarify the relationship between many variables in the environment, a **phase wheel** is plotted with the 'spokes' showing the relative timing of the changes for each factor (Clemens and Prell, 1990; Imbrie *et al.*, 1992). A separate phase wheel is required for each scale of cyclicity.

4.6   Wavelet analysis

In Fourier analysis the sine and cosine functions, used for the Fourier transform, extend along the whole length of the time series. The result is that the spectrum, for example, indicates average power. Evolutionary spectra can be used to study gradual changes in the data (Section 4.2). However, for non-stationary data, abrupt changes in the variance distribution are a problem. This is because of the trade-off between frequency resolution and time/stratigraphic resolution in evolutionary spectra.

An alternative approach to non-stationary data is to use **wavelet analysis** (Hubbard, 1996; Mallat, 1998; Torrence and Compo, 1998). Wavelets are orthogonal (i.e. independent) functions where the oscillations die away to zero after a short interval rather than going on indefinitely (Fig. 4.17). A wavelet can be used for multiplication with a time series to produce a wavelet transform, analogous to the use of sine and cosine waves for the windowed Fourier transform in evolutionary spectra. Although there are restrictions in the permitted characteristics of wavelets, the different types exhibit a huge variation in shape (Mallat, 1998). For each application a particular shape, or '**mother wavelet**' is chosen. For analysing the non-stationary character of a time series, the mother wavelet is varied in wavelength (stretched or compressed,

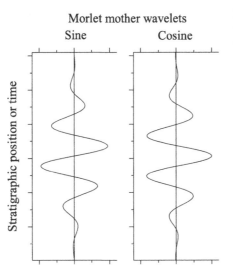

Morlet mother wavelets

Sine   Cosine

Stratigraphic position or time

Fig. 4.17 Morlet mother wavelets with five oscillations.

so-called '**dilation**') as well as used in different stratigraphic positions or time periods ('**translation**').

For rapid computation the **fast wavelet transform** is available for a restricted set of discrete time wavelet types (Hubbard, 1996). In the fast wavelet transform, first the time series is divided, by filtering, into the low-frequency components and the high-frequency components. The high-frequency component series is then multiplied by the smallest version of the mother wavelet after successive shifts in the wavelet position along the length of the data. The remaining low-frequency component series is then divided again into low- and high-frequency parts and the wavelet transform is re-applied to the high-frequency component, but using a wavelet wavelength which is double the previous scale. This is repeated until the very lowest frequency components of the original time series have been transformed (Hubbard, 1996).

This procedure can be understood by reference to **tiling** or an examination of the parts of a time–frequency plot covered by wavelets with different scales (Fig. 4.18). In Fourier spectra, or evolutionary spectra, the same length of data is considered regardless of frequency and the frequency resolution is constant across the spectrum. However, in wavelet tiling long wavelength wavelets cover a large proportion of the data. Consequently the frequency resolution is good, but the time or stratigraphic resolution (i.e. the ability to monitor non-stationary behaviour) is poor. On the other hand, short wavelength wavelets require fewer data so the time or stratigraphic resolution is good, but the frequency resolution is poor. This trade-off between time resolution and frequency resolution is described as an uncertainty principle (Mallat, 1998). The factor of two involved in successive frequency estimates for the fast wavelet transform

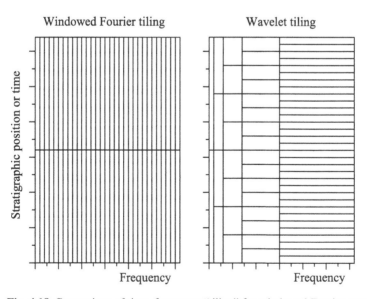

Fig. 4.18 Comparison of time–frequency "tiling" for windowed Fourier spectra and a wavelet spectrum. Note that in both cases it is possible to use an overlap of the tiles rather than the non-overlapping tiling illustrated.

means that wavelet spectra are usually illustrated with the frequency axis plotted in base two logarithms (Fig. 4.19).

Many types of individual mother wavelets have oscillations that vary in wavelength. However, Morlet wavelets (Fig. 4.17) have a constant wavelength and therefore have been popular in cyclostratigraphic work because the frequency significance of the output is obvious (Kumar and Foufoula-Georgiou, 1994; Prokoph and Barthelmes, 1996; Yiou *et al.*, 1996; Berger *et al.*, 1997). These studies have mainly used wavelets

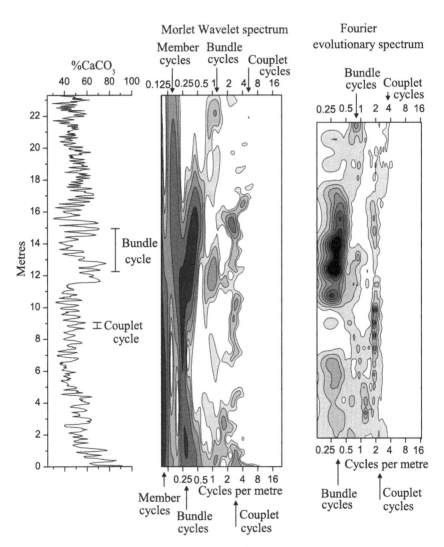

Fig. 4.19 Comparison of a Morlet wavelet spectrum with a Fourier multi-taper method evolutionary spectrum (from Fig. 4.2). Note the base-two log frequency scales nece-ssitated by the wavelet tiling (Fig. 4.18). 'Member cycles' refers to the approximately 12-metre-scale bundle of bundles as described by Weedon and Jenkyns (1999). Note that the wavelet spectrum covers a wider range of frequencies than the Fourier evolutionary spectrum.

to provide a form of high-resolution evolutionary spectrum. Torrence and Compo (1998) discuss tests of statistical significance and the way that the ends of the data affect the spectrum within the '**cone of influence**'.

Currently the application of wavelet analysis has been fairly restricted, perhaps partly because non-stationary series have been avoided by cyclostratigraphers due to the difficulty of demonstrating regular cyclicity throughout the record. Other applications of wavelets are available. For example, in oceanography wavelet versions of evolutionary coherency spectra and phase spectra have been used to examine the initiation of water waves under the influence of variable winds (Liu, 1994).

4.7   Phase portraits and chaos

One procedure frequently used for studying time series in the context of non-linear systems is to reconstruct the **phase portrait**. Phase portraits reveal how the output from a system, the time series, evolves(d) geometrically in **phase space**. In some cases inspection of the phase portrait helps interpret the system that generated the time series. For systems described by known equations relating to continuous time, it is relatively straightforward to reconstruct the phase portrait. However, for real cyclostratigraphic time series (with discrete time, a single realization and a limited number of data points) one does not know the equations that governed the environmental system. This means that reconstruction of the phase portraits usually requires the **time delay method** (Williams, 1997). In this procedure one simply plots each time series point against another point at a different, fixed, offset (i.e. at a different lag). For example, one could plot each value (x_i) versus the previous value (x_{i-1}) in which case the offset or **delay time** is one. This would produce a plot with two axes – it would be two-dimensional. The number of axes used is called the **embedding dimension**. For example, three embedding dimensions could use axes with x_i versus x_{i-1} versus x_{i-2} (i.e. a delay time of one), or perhaps x_i versus x_{i-4} versus x_{i-7} (a delay time of three). There is no simple way to determine the appropriate embedding dimension (Kantz and Schreiber, 1997), but when it exceeds three the resulting phase portrait cannot be plotted clearly on paper. An alternative to the time delay method is to plot each phase space axis using a different measured variable (e.g. Saltzman and Verbitsky, 1994; Williams, 1997). Gipp (2001) illustrates examples of the use of the **time derivative method** where a variable is plotted against its derivative(s).

To gain an understanding of phase portraits consider Fig. 4.20. The Earth's obliquity or tilt of the equator relative to the plane of the orbit, the ecliptic, varies from about 22° to 24.5° over periods of close to 41 ka (Section 6.9.1c). A phase portrait of successive values reveals a tight spiral elongated at 45° to the axes. This indicates that successive values are closely related, or autocorrelated, at a lag of 1. To get a better indication of the portrait a larger time delay is also plotted and this more clearly shows the spiral **trajectory**. It should be clear that if a perfectly periodic cycle (in non-linear systems terminology, a '**limit cycle**') of constant amplitude had been plotted, the trajectory

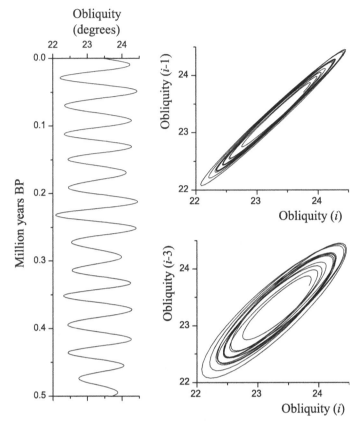

Fig. 4.20 The appearance of the phase portrait of orbital obliquity from the last half million years depends on the embedding dimension. When successive values are plotted against each other the strong autocorrelation generates a narrow band of cycling values (top right). With a longer delay between plotted values the individual cycles are more apparent (bottom right). BP: before present.

would have been a simple ellipse. However, the orbital tilt cycle is quasi-periodic so the trajectory never quite retraces its earlier path.

To allow the trajectory to be visualized, and since there are relatively few data values, the figures in this section have phase portraits with straight lines connecting the points. However, it is more usual to plot the points alone. If using points instead of lines, and in the case of the Earth's obliquity, the plot using a delay time of one would have consisted of an elongate cloud of points clearly indicating the autocorrelation. However, with a three-dimensional plot quasi-periodic oscillations produce points distributed on the surface of a torus (or an American donut shape).

The phase portrait for the Earth's precession cycle (Section 6.9.1a), with a period close to 21 ka, is more complicated. The amplitude varies substantially through time so that the spiral trajectory covers a much wider area of phase space than the obliquity trajectory (Fig. 4.21). In addition, the centre of the spiral moves slightly through time

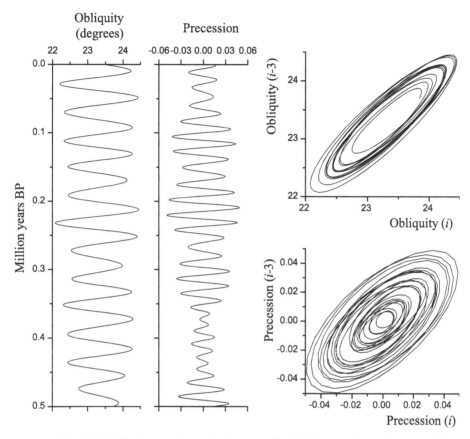

Fig. 4.21 Orbital precession variations are relatively large (due to the eccentricity-related heterodyne amplitude modulation) compared to obliquity variations. Consequently the phase portrait has values occupying the centre of the plot as well as the margins.

because the mean precession varies slightly. Precession exhibits heterodyne amplitude modulation (Section 1.3) linked to the Earth's eccentricity cycles (Section 6.9.1b). The changing mean precession is linked to the 100 ka and 400 ka eccentricity periods. If the mean precession changed more substantially the spiral would move more widely over the phase space. Potentially then the presence of more than one quasi-periodic component in a time series would produce a trajectory that corresponded to a spiral moving back and forth through time, making it very difficult to see the basic spiral pattern on a static plot.

Similarly if the time series is non-stationary in the mean and/or variance (mean amplitude) the phase portrait is much harder to interpret. Additionally, all observational time series also have noise components. The noise, arising from factors such as measurement errors, means that the trajectory is no longer smooth so the plot jumps erratically around the expected pathway. If the quasi-periodic components have spectral peaks that are small relative to the continuum noise, the noise will cause the whole volume of the phase space to become filled and no simple trajectory can be observed regardless of the number of embedding dimensions.

It has been demonstrated previously that the lower two-thirds of the Belemnite Marls carbonate data contain quasi-periodic components and there is certainly noise present (Section 4.2). The phase portrait of the detrended data, using an embedding dimension of two, looks fairly erratic (Fig. 4.22). There is a hint of spiral structure due to the decimetre-scale couplet cycles, but the noise has created a rather ragged appearance. Using three embedding dimensions it is easier to see the spiral trajectory – indicating that the trajectory is not a simple flat spiral. Note that the metre-scale bundle cycles will have caused the spiral related to the couplet cycles to move back and forth across the phase space as discussed for the Earth's precession history.

To complicate matters it is now clear that the continuum noise in power spectra can partly arise from systems exhibiting **deterministic chaos**, a term used in the

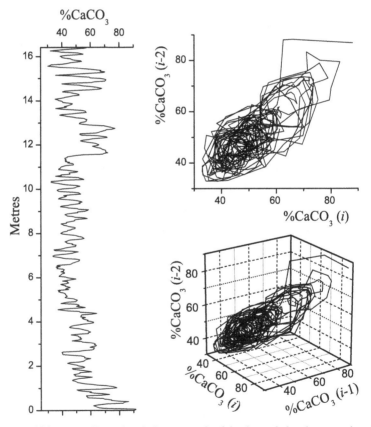

Fig. 4.22 A two-dimensional phase portrait of the detrended carbonate values from the lower two-thirds of the Belemnite Marls (top right) has a rather ragged appearance. This reflects the presence of analytical noise, the large amplitude variation and the presence of the bundle-scale variations in the mean carbonate content of successive couplet cycles. Plotting using three axes, or using three embedding dimensions (bottom right), leads to a clearer picture of the geometry in this case.

field of non-linear dynamics. If chaos is present then the phase portrait can exhibit a definite, though complex, structure as explained below. A very brief outline of some of the issues is provided here. For those interested in the conceptual importance of **chaos theory**, Stewart (1990) is an excellent starting point. The mathematics of non-linear dynamics and chaos has been treated in a straightforward, and highly readable, manner by Williams (1997). For more complex issues with a firmly practical, rather than just a theoretical, approach consult Kantz and Schreiber (1997). Turcotte (1997) and Middleton *et al.* (1995) discuss chaos theory and wider time series issues concerned with non-linear dynamics.

Standard linear systems theory had implied that to get an unpredictable, noisy output one needs a noisy input. Stochastic noise, and the spectral continuum, of cyclostratigraphic data can certainly be explained as the product of a combination of factors such as measurement errors, interpolation, bioturbation, accumulation rate variations and undetected hiatuses (Chapter 5). Yet chaos theory has established that, in the right conditions, certain relatively simple non-linear equations can generate time series that have a continuum spectrum. The output is indistinguishable from noise using standard time series methods because it is impossible to predict successive values unless the initial conditions are precisely known. This is despite the fact that every value in the output is strictly deterministic (the product of an equation).

To see why prediction is impossible, consider an equation designed to generate discrete time output, the tent map. This can be written as $x_i = k[1 - \mathrm{ABS}(1 - 2x_{i-1})]$, where x_i is the new value, x_{i-1} is the previous value and ABS means 'take the absolute value' (i.e. ignore the sign). This very simple equation produces chaotic output when k lies anywhere between 0.5 and 1.0 (Turcotte, 1997). Figure 4.23 illustrates 30 successive values from the tent map using $k = 0.990$ and two different starting values. It is clear that the values in the two outputs diverge after the eighth time step. The starting values differed by just 0.2% so the output depends critically on the starting point. This illustrates **sensitive dependence on initial conditions**, which is characteristic of chaos. In fact even an infinitesimal difference in starting values will result in diverging output. As a consequence, without knowing the equation and exact starting conditions it is impossible to predict the future values after just a few time steps. Chaos theory partly developed out of meteorological studies where it has become clear that despite attempts to model the weather and establish starting conditions, all predictions are unreliable after a few days, or exceptionally, a week or two (Stewart, 1990). In chaotic systems the length of time before predictions are unreliable (i.e. output values diverge substantially) depends on the state of the system (or the value of k in the example of the tent map). It is disputed whether the climate and climatic records are actually chaotic (Nicolis and Nicolis, 1984; Grassberger, 1986; Mudelsee and Stattegger, 1994; Richards, 1994; Gipp, 2001). However, certain relatively simple non-linear numerical models of climate generate time series with red noise spectral backgrounds and superimposed quasi-periodic components related to the annual and Milankovitch orbital cycles (Le Treut and Ghil, 1983; James and James, 1989).

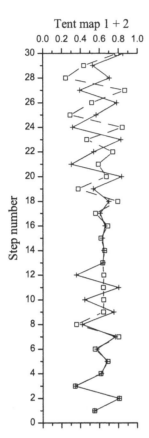

Tent map 1 + 2

Fig. 4.23 Even though successive values are determined via equations, pairs of chaotic time series rapidly diverge if the starting conditions are not identical. In this case output from the tent map, with starting values that differ by just 0.2%, diverge significantly after just eight steps. This illustrates sensitive dependence on initial conditions.

The tent map generates time series with a continuous **blue-noise** spectral background (i.e. dominant high frequencies, Fig. 4.24). As red noise is more typical of cyclostratigraphic data, the tent map data in the figure have been smoothed using one application of the Hanning weights (0.25, 0.5, 0.25). This is merely to indicate that even a chaotic system generating blue noise may be recorded stratigraphically as red noise following processes such as bioturbation, etc. (Section 5.3.2). The phase portrait of the smoothed tent map output looks ragged with two axes, but with three, the characteristic shape is obvious (Fig. 4.25). In fact all chaotic records normally require three axes to be able to visualize the shape of the trajectory (Williams, 1997).

The shape of chaotic phase portraits is highly complex and is described as having a '**strange attractor**' in recognition that the trajectory eventually approaches (is 'attracted' to) a special geometry. The term attractor refers to the trajectory followed by the system after a number of transient values that do not lie on the attractor itself. The special geometry mentioned is fractal (similar in appearance at different scales of observation) meaning that it does not form a simple solid shape, such as an ellipsoid, in three dimensions. Instead it has a non-integer dimension as explained by Williams (1997). However, though characteristic, a non-integer dimension is not diagnostic of

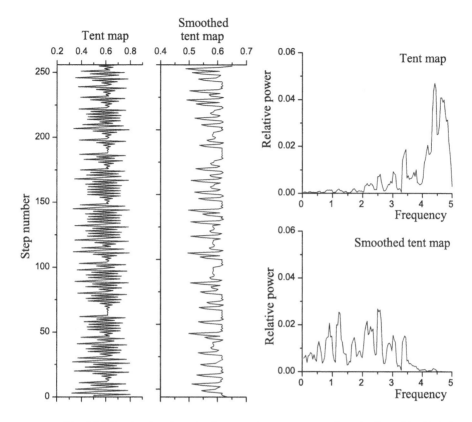

Fig. 4.24 Purely chaotic series have continuum spectra often dominated by high frequencies. In the case of the tent map the spectrum corresponds to blue noise. Even minimal smoothing of such a record (e.g. by bioturbation of sediments) generates a reddish spectrum.

chaos since there are systems with a non-integer dimension that do not show sensitive dependence on initial conditions (Medio, 1992).

One way to demonstrate the presence of chaos quantitatively is to obtain the largest **Lyaponov characteristic exponent (LCE)**. Consider a trajectory in phase space moving round repeatedly. Choose an initial point on the trajectory and follow the pathway a short way – this represents the reference trajectory. Next choose another point on a part of the whole trajectory that is passing very close to the initial point, but represents a different time interval, and determine the separation of the second length of trajectory from the reference. If the two pathways separate at an exponential rate then the rate is positive and known as a local Lyaponov exponent. Exponentially diverging trajectories are indicative of chaos as they indicate sensitive dependence on initial conditions (Fig. 4.23). However, different parts of a strange attractor can yield different exponents. Therefore, to establish chaos one needs to repeat the exercise for all parts of the attractor. Hence, the largest, or maximal, Lyaponov exponent indicates the average rate of separation of nearby trajectories for the whole of the attractor reconstructed in phase space. If the value is positive, but not infinite, then there is sensitive dependence

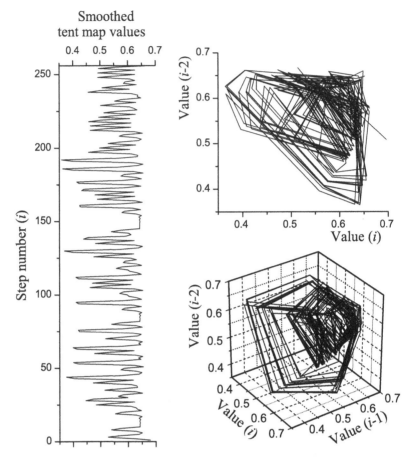

Fig. 4.25 The phase portrait of the tent map using two embedding dimensions is insufficient to resolve the structure (top right). With a three-dimensional plot it is possible to identify the characteristic attractor geometry (bottom right).

on initial conditions and chaos present. However, this method requires a large number of movements around the phase space containing the reconstructed attractor. This is so that there are many nearby sections of trajectory for determining local Lyaponov exponents. Depending on the nature of the system being studied and the method used to determine the largest Lyaponov exponent, between tens of thousands and a million data points are required to get reliable results (Kantz and Schreiber, 1997).

Despite the relative youth of the discipline, chaos theory is already pervasive in the physical sciences. However, it has yet to make a big impact in cyclostratigraphic time-series analysis despite its undoubted conceptual importance (Smith, 1994). This lack of impact arises for several reasons. Most obviously cyclostratigraphic time series are generally far too short for standard testing for the presence of chaos. Nevertheless, there are now methods that potentially can be used in a qualitative fashion to indicate chaos or at least non-linear dynamics in short records (Sugihara and May, 1990; Wales, 1991; Tsonis and Elsner, 1992; Barahona and Poon, 1996).

Chaos analytical methods require time series that are strictly stationary (i.e. stationary in more factors than just the mean and variance, Kantz and Schreiber, 1997). Since cyclostratigraphic data exhibit a red noise background virtually without exception (Section 6.2), the longest wavelength/period components are repeated too infrequently for meaningful inferences to be obtained (Kantz and Schreiber, 1997). The red noise character is indicative of autocorrelation (correlation of successive values). As shown in Fig. 4.20, autocorrelation means that the trajectory is not sufficiently separated in phase space. This can be overcome either by appropriate pre-processing of the data (e.g. pre-whitening) or by using a phase space with a suitably large delay time (Kantz and Schreiber, 1997; Williams, 1997). Note that interpolated data inevitably have a significant autocorrelation, so it is better to obtain equally spaced data if possible. Finally, for the most meaningful analysis for chaos, the data need to be pre-processed to remove the trend and the periodic and quasi-periodic components (Williams, 1997). This leaves a record potentially dominated by stochastic noise (including measurement errors) which may obscure any chaotic signal present. Not surprisingly then, firm demonstration of chaos has only been widely accepted in the physical sciences for time series obtained from carefully designed experimental systems where factors such as analytical noise can be controlled and large numbers of data points obtained. Further difficulties in using real data sets for studying non-linear dynamics and chaos are discussed in Chapters 5 to 7 of Kantz and Schreiber (1997).

Recently the concept of **self-organized criticality** has been introduced. Chaotic systems involving large numbers of degrees of freedom or dimensions exhibit self-organized criticality. In such systems the size and frequency of events follows a power law (Section 3.3.1). Thus the larger an event the rarer it is, but the size of each individual event cannot be predicted. Systems that exhibit self-organized criticality are said to be 'marginally stable' (Turcotte, 1997). In other words, a tiny disturbance can trigger an event, big or small, the outcome of which restores the system to the marginally stable state. This concept has been applied to situations as diverse as the sizes and frequencies of earthquakes to the variability of stock market indices (e.g. Turcotte, 1997; Buchanan, 2000).

4.8   Singular spectrum analysis

Singular spectrum analysis provides a way to analyse even small data sets (e.g. less than 150 points) in a non-parametric manner to extract the key features (Vautard and Ghil, 1989). It is closely allied to principal component analysis (Davis, 1986). The first step is to consider the phase space representation of the data using a large embedding dimension (usually less than a third of the data points). From the previous section it will be apparent that whatever shape the phase portrait has, it is likely to be complex and may require many axes to bring out the full variability. For example, for the Belemnite Marls data there are certain directions in phase space which involve a greater range or variability/variance than others (Fig. 4.22). Singular spectrum analysis is able to find the directions in phase space that involve the greatest variability.

The data are then treated as a series of variables, one for each embedding dimension (or lag). Each lagged version of the data (variable) is associated with a variance, plus the covariances with every other variable. (Covariance is just the correlation coefficient multiplied by the product of the standard deviations of each variable, Davis, 1986.) A matrix is formed of the variances and covariances for every lag being considered. For example, with an embedding dimension of two (which would not be suitable for singular spectrum analysis), the first variable X would be x_i and the second variable, or Y, could be x_{i-1}. The first row of the matrix would be the variance of X and the covariance of $X + Y$, and the second row would be the covariance of $X + Y$ and the variance of Y. Since the matrix of variances and covariances is symmetrical it yields real eigenvalues and real eigenvectors (i.e. values that do not involve complex numbers, eigenvalues and eigenvectors are explained below, Davis, 1986).

Using the two-dimensional example, the two rows of the matrix could be used as coordinates for two points in phase space. The two points can be used to define an ellipse centred on the origin. The eigenvalues correspond to the length of the long and short axes of the ellipse. Similarly the eigenvectors define the slopes, and hence the orientation in phase space, of the two axes (Davis, 1986). In a real situation one would be faced with many eigenvectors – reflecting the many embedding dimensions of the phase space. In singular spectrum analysis the eigenvector axes are orthogonal to each other so that the variability along one axis is independent of that in any other.

The plot of the square root of the eigenvalues (i.e. the singular values – hence the synonym **singular value decomposition**), ranked according to size, is called the singular spectrum (Vautard and Ghil, 1989). It is usual to plot the log of the singular values so that small values can be compared with large. In Fig. 4.26 the singular spectrum of the Belemnite Marls data shows, as for all cyclostratigraphic data, a series of decreasing values. Looking at the uncertainty in these values (the 95% confidence interval) it is clear that, for example, the first as well as the second, third and fourth eigenvalues are statistically distinct from (larger than) the others.

The clever part of singular spectrum analysis is in the isolation of the independently varying components. The eigenvalues and eigenvectors define **empirical orthogonal functions** (**EOF**s) in phase or lag space (Vautard and Ghil, 1989). These can be projected to yield the principal components as a function of time or stratigraphic position (Fig. 4.27, e.g. Dunbar *et al.*, 1994). The sum of the power spectra of the individual principal components corresponds to the power spectrum of the original time series (Vautard and Ghil, 1989).

Typically individual principal components are dominated by particular frequencies. For example, in Fig. 4.27 principal components (PCs) 2 and 3 have the wavelength of the decimetre couplet cycles. Note that these components reveal the same amplitude variations – they form a natural pair since they represent cosine and sine counterparts (they are described as existing in **quadrature** or having a 90° phase difference). However, though PC 1 predominantly consists of oscillations related to the metre-scale bundles, also present is a small-amplitude oscillation related to the couplets. The figure shows the result of adding, point-by-point, the first three principal components,

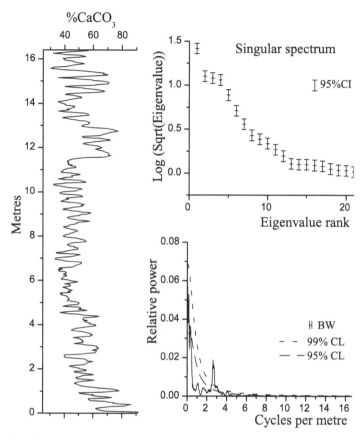

Fig. 4.26 Usually the singular spectrum is plotted using the log of the singular values (i.e. the log of the square root of the eigenvalues) ranked using decreasing variance (i.e. decreasing singular value). The vertical bars indicate the 95% confidence intervals of the singular values. For the log version of the singular spectrum the uncertainty, or the 95% confidence interval, is $\pm(1.96/N)^{0.5}$ where N is the number of data points (p. 408 of Vautard and Ghil, 1989). Sqrt: square root.

which in this case account for 50.8% of the variance (or 71.3% of the amplitude) of the original data. The outcome is clearly a filtered version of the original data. Thus singular spectrum analysis can be used as a rapid method for FIR filtering – for example, high-frequency noise can be removed (Kantz and Schreiber, 1997). However, since individual PCs can contain oscillations of more than one frequency, the results are not optimal and should always be checked using a power spectrum of the added principal components.

The procedure for generating a filtered version of the Belemnite Marls data just outlined corresponds broadly to the original use of singular spectrum analysis (e.g. Schlesinger and Ramankutty, 1994; Plaut *et al.*, 1995). It assumes that it is sufficient to identify individual eigenvalues or pairs of eigenvalues that are distinct from the others in terms of the 95% confidence interval (Vautard and Ghil, 1989). However, for cyclostratigraphic data, one really needs to compare the calculated singular

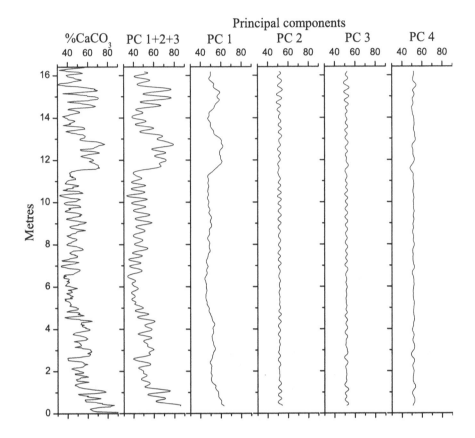

Fig. 4.27 The first four principal components of the Belemnite Marls data derived from the eigenvectors identified by singular spectrum analysis in Fig. 4.26. All plots use the same vertical and horizontal scales. Note that the singular spectrum analysis was conducted using linearly detrended data. The first three principal components alone account for about 50% of the total variance of the series.

spectrum with the results that would be obtained from analysis of first-order autoregressive noise. For example, Fig. 4.28 illustrates the singular spectrum for AR-1 noise with the same number of points and the same lag-1 autocorrelation as the data from the Belemnite Marls. The result is a singular spectrum with many of the characteristics of Fig. 4.26. Hence, Monte Carlo singular spectrum analysis has been introduced (Allen and Smith, 1996). This requires the generation of thousands of simulations or individual AR(1) realizations with the same broad statistics as the original data. The observed singular spectrum can then be compared with the ensemble results of the Monte Carlo simulations (e.g. Biondi *et al.*, 1997). Allen and Smith (1996) demonstrated that certain earlier results using the original procedures could not be supported. Finally, another development of singular spectrum analysis has been a combination with wavelet analysis to allow examination of non-stationary data (Yiou *et al.*, 2000).

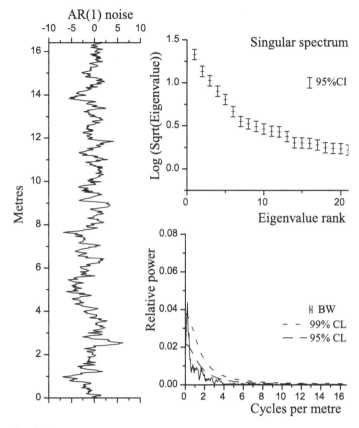

Fig. 4.28 First-order autoregressive noise generated using the same number of points and the same lag-1 autocorrelation as the Belemnite Marls data in Fig. 4.26. The noise values have been treated as though they were collected at the same sample interval as the real data (i.e. 0.03 m) and the embedding dimension used for the singular spectrum analysis (20) is also the same. All the plots use identical scales to Fig. 4.26 for ease of comparison. Only the first eigenvalue is statistically distinguishable from the rest. Sqrt: Square root.

4.9   Chapter overview

▇ For long time series (e.g. more than 20 times the length of the shortest regular cycle) non-stationary variance can be investigated using evolutionary spectra (evolutive or windowed Fourier analysis), or using wavelet spectra.

▇ Filtering can be used to manipulate time series to isolate the high frequencies, the low frequencies or a narrow band of frequencies. When band-pass filtering, prior spectral analysis is needed to demonstrate the presence of a significant concentration of variance (a spectral peak) at the appropriate frequencies.

▇ The history of amplitude and phase variations of a regular cycle can be determined using complex demodulation.

▨ Pairs of time series (i.e. records of two variables) can be compared using coherency and phase spectra. Significant coherency indicates similar amplitude
variations and thus the possibility of a causal connection in the environment
between the two variables at a particular frequency. Phase spectra can reveal the
relative timing of variations in the two time series with reasonable certainty, at
frequencies where the coherency is significant.

▨ A time series in phase space (the data plotted against itself at different lags)
provides a context for phase portraits and singular spectrum analysis (SSA).
SSA can sometimes be used in a manner analogous to spectral analysis and
for filtering. Non-linear time-series analysis involves treatment of data in phase
space, but requires strict stationarity and tens of thousands of data points (both
are rare or non-existent in cyclostratigraphic data sets).

Chapter 5

Practical considerations

5.1 Introduction

This chapter is concerned with various practical matters not considered elsewhere. Ideally the various types of environmental cyclicity discussed in Chapter 6 would produce cyclostratigraphic time series that faithfully reproduce the original temporal information. Unfortunately this is never the case because of a variety of factors discussed below. Cyclostratigraphic signals are invariably distorted records of the environmental variables due to modification of both the amplitude and frequency characteristics. It is important to be aware of these distortions, both so that interpretations are not pushed too far and so that, in some cases, remedial action can be taken. The distortions tend to reduce the chances of distinguishing spectral peaks from the spectral background. Additionally, sometimes the distortions introduce additional, often easily identifiable, spectral components that describe the non-sinusoidal shapes of time-series oscillations.

Once the distorting processes have been covered, the chapter continues with practical issues which are, at first sight, straightforward. These concern the number of data points needed for a meaningful time-series analysis, regularity and the interpretation of spectral peaks.

In order to illustrate the effects of distortion processes the same 'test series' has been used in several figures. This series consists of a 4 m sine wave plus a 0.75 m sine wave, both of constant amplitude, plus AR(1) noise (with the lag -1 autocorrelation $= 0.7$). The figures allow comparison of depth records that are undistorted, with depth records that result from the various distortions. The distortion processes have been simulated fairly crudely, but with enough similarity to real distortions to gauge, qualitatively, the effects discussed.

5.2   Cyclostratigraphic signal distortions related to accumulation rate

Linear sediment accumulation rate, as used here, refers to the thickness of sediment deposited per unit time interval after final burial (i.e. after hiatuses related to the environment of deposition have been created). Throughout the book the abbreviation 'accumulation rate' means linear sediment accumulation rate and *not* mass accumulation rate. Average sediment accumulation rate depends upon the type of sediment being deposited, the environment of deposition, compaction and the completeness of the section – which is linked to the time span of measurement (e.g. Sadler, 1981).

There are essentially four ways in which variations in accumulation rates can distort the relationship between time and sediment thickness. Thus accumulation rates might exhibit: a trend (i.e. a predominantly monotonic increase or decrease); random variations; an abrupt or stepped change; or variations that are partly determined by the environmental signal of interest (Fig. 5.1).

5.2.1   Trends in accumulation rate

A progressive increase (or decrease) in accumulation rate will cause regular cycles in time to be encoded in strata as successively thicker (or thinner) oscillations. Such changes in mean rate reflect environmental variations that have periods longer than the length of the cyclostratigraphic record. Examples might include million-year or tens-of-million-year variations in relative sea level, changes in global ocean circulation

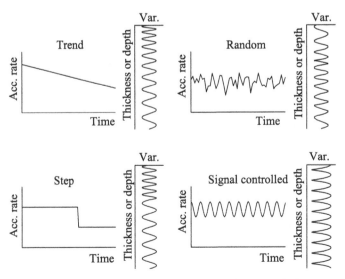

Fig. 5.1 Four different ways in which variations in accumulation rate might vary and affect a time series. In real cases several of these types of variation may be combined. Acc. rate: accumulation rate; Var: variable.

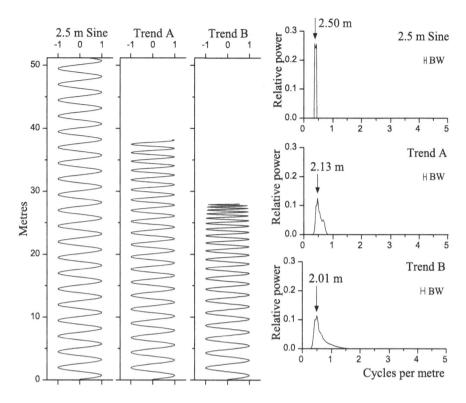

Fig. 5.2 A trend in accumulation rate has the effect of spreading the power associated with the original regular oscillations across a range of frequencies. As a result the peak height decreases. The frequency of the peak maximum also shifts, to a position between the frequencies corresponding to the longest and shortest wavelength oscillations.

patterns related to climate, or processes connected to tectonic plate movements and orogeny. The data will not be stationary in terms of the variance–frequency relationship in stratigraphic data where there is a trend in accumulation rates, but such a trend might not be apparent from visual inspection.

Figure 5.2 illustrates the effect on a power spectrum of progressively reducing the wavelength of the oscillations from 2.5 to 1.25 m. The narrow spectral peak of the undisturbed series is replaced by a broader, and smaller, peak for the series with a trend in accumulation rates. This simply indicates that the spectral peak associated with the trend relates to oscillations that have a broad rather than a narrow range of frequencies. With a more rapid change in mean accumulation rate, relative to the length of the undisturbed cycles, the peak becomes even broader and smaller.

Clearly if accumulation rates change 'too quickly' the associated spectral peak is likely to be so broad and small that it would be indistinguishable from background spectral noise. This scenario is shown in Fig. 5.3, where the trends of the previous figure have been applied to the test series. The figure shows that high-frequency peaks become broader and smaller more quickly than low-frequency peaks. In the case of the more

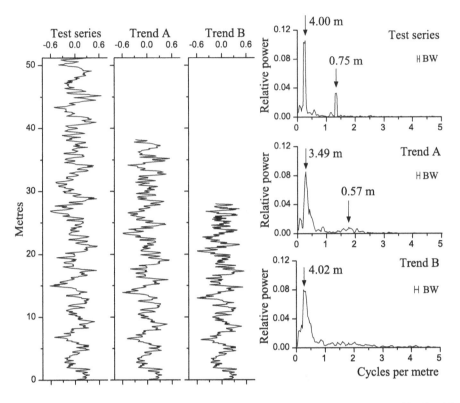

Fig. 5.3 Using the test series and the trends in accumulation rates employed in Fig. 5.2. Note that the combination of background noise with a stronger and stronger trend makes the high-frequency peak indistinguishable from noise before the low-frequency peak becomes indistinguishable. This reflects the higher wavelength resolution of the spectrum at higher frequencies. Hence, a fixed proportionate variation in wavelength spreads the power over a wider range of frequencies at the higher frequencies.

severe trend, the oscillations that originally had a wavelength of 0.75 m no longer generate a spectral peak. This effect results from the greater wavelength resolution of the spectrum at high frequencies (following from the constant frequency resolution). This means that the same proportionate change in wavelength spreads out the power for high-frequency oscillations over a wider frequency band than for low-frequency oscillations.

In real depth records, trends in accumulation rates can often be recognized by visual inspection, especially if there are just a few dominant components present. It is difficult to remove trends without knowing something about the generating processes (Schwarzacher, 1975). Fourier evolutionary spectra and wavelet analysis can be used to identify components that have a gradually changing frequency (Sections 4.2 and 4.6). These methods might reveal the wavelength of particular components in successive stratigraphic intervals. If it can be inferred that the components relate to environmental cycles with a constant, or near constant period, then the changing wavelengths in the depth records reveal the changing accumulation rates – and potentially the trend could be removed from the record.

5.2.2   Random changes in accumulation rate

Environmental variables are naturally noisy, even if there are periodic or quasi-periodic components. Consequently accumulation rates are unlikely to remain constant through time, even if the mean remains unchanged (i.e. trend-free). Clearly random variations in accumulation rates have the potential to cause variations in the thickness of sedimentary cycles that are primarily driven by regular or periodic processes. This problem is well known (e.g. Pestiaux and Berger, 1984b; Schwarzacher, 1991, 1993; Herbert, 1994; Meyers *et al.*, 2001). The consequences are that, as for trends in accumulation rates, there is a broadening and decrease in size of spectral peaks related to the regular cycles (Fig. 5.4). As mentioned before, the greater the variation in accumulation rates, the more the distortion of the spectrum, and again expected high-frequency spectral peaks will become indistinguishable from noise before low-frequency peaks.

Note that in the test series example (Fig. 5.5), the broadening of the high-frequency peak results in power being added to what, in the original signal, was a background noise peak. As a result in the second distorted record the high-frequency peak appears to

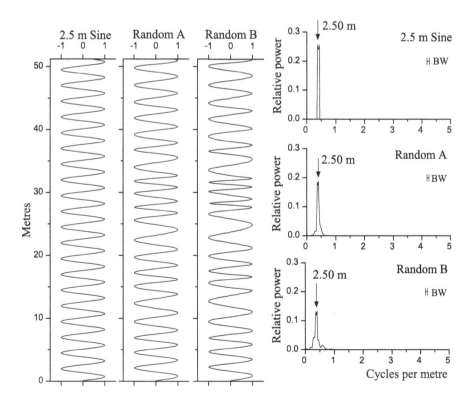

Fig. 5.4 Random variations in accumulation rates lead to broadening of spectral peaks in a manner similar to trends in accumulation rates. However, the position of the spectral peak in this case remains unchanged.

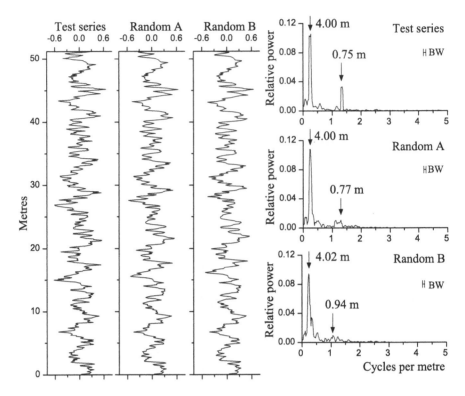

Fig. 5.5 The test series distorted using the same random variations in accumulation rates applied in Fig. 5.4. The more variable the variations in accumulation rates, the more likely that high-frequency spectral peaks become indistinguishable from the noise background (cf. Fig. 5.3).

have moved to a different frequency (1/0.94 m). This implies that random accumulation rates have the potential to modify apparent wavelength ratios – which have often been used to identify the type of regular environmental process involved. How likely wavelength-ratio modification is requires investigation. Nevertheless, it appears to depend on how much the original peaks have been broadened, and whether there are nearby noise peaks in the spectrum of the environmental variable.

Since random variations in accumulation rate can severely distort the spectrum it is desirable to remove their effect when possible. This is the case for Neogene to Recent deep-sea sediments when orbital cycle ('astronomical') time scales are of interest. The procedure of orbital tuning matches records of varying sediment composition or isotopic signals to either the calculated orbital-cycle history or models of global ice-volume history, as driven by the orbital cycles (Section 6.9.3). The orbital tuning allows the history of accumulation rate variations to be established between the control points. In the absence of a tuning target, graphic correlation (Shaw plots) can be used to compare adjacent stratigraphic records. Shaw plots provide an indication of the relative changes in accumulation rates between localities or with respect to a standard section (e.g. Prell *et al.*, 1986).

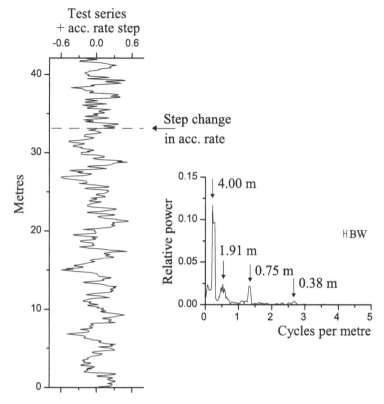

Fig. 5.6 A step change in accumulation rates applied to the test series. The two accumulation rates lead to a doubling of the number of significant spectral peaks (i.e. peaks distinguishable from the noise). acc. rate: accumulation rate.

5.2.3   Abrupt or step changes in accumulation rate

Abrupt changes in accumulation rates reflect either abrupt changes in the environment of deposition, or hiatuses. Obviously time series incorporating a single step change will generate twice the number of spectral peaks as a step-free record (unless, by chance, some peaks are 'shifted' to the frequencies of peaks detected in the other section, Fig. 5.6). When there are multiple steps the spectrum will contain a plethora of large peaks. In such cases it may only be analysis of step-free records above, below or lateral to the distorted record that indicates a problem. Wavelet analysis is particularly good at identifying this type of non-stationarity (Prokoph and Barthelmes, 1996).

5.2.4   Signal-driven accumulation rates, harmonics and combination tones

Conceivably variations in accumulation rates may be linked to the environmental variable recorded by the cyclostratigraphic time series (Fig. 5.1, Schiffelbein and Dorman, 1986). In this case the time series observed as a function of depth or thickness will be distorted, in a non-linear manner, relative to the time domain. Weakly non-linear systems (as opposed to strongly non-linear systems that may be chaotic, Section 4.7) characteristically generate distortions in time series that result in spectral components called **harmonics** and **combination tones** (e.g. King, 1996). Sometimes the environmental variables themselves involve harmonics or non-sinusoidal variations in time (e.g. Short *et al.*, 1991; Hagelberg *et al.*, 1994; McIntyre and Molfino, 1996; Berger and Loutre, 1997).

For example, if a single sinusoid is progressively distorted in (linear or non-linear) proportion to the sine wave amplitude, the resulting record has shorter troughs than peaks or shorter peaks than troughs. Schiffelbein and Dorman (1986) used an equation for producing varying degrees of linear distortion, i.e. the distortion of the output depth scale depended linearly on the input. A non-linear distortion could involve a depth scale that depends on some power of the input values. In either case, the result is that the spectrum of the time series in the depth domain will contain harmonic peaks that occur at integer multiples of the frequency of the primary peak, or **fundamental**. So if the fundamental has a frequency f, the harmonic peaks appear at $2f$, $3f$, $4f$, etc. (Fig. 5.7). Some authors refer to the harmonics with frequencies of $2f$, $3f$, $4f$, etc.

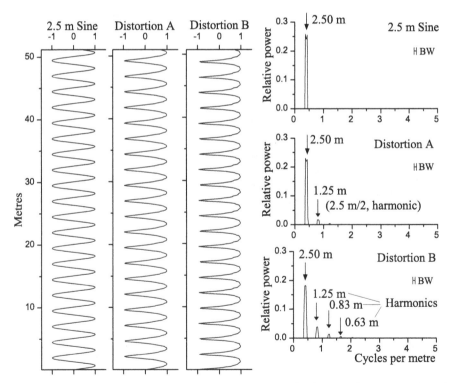

Fig. 5.7 The change from sinusoidal to cuspate oscillations results in harmonic peaks in a spectrum. In this case the distortions result in horizons with large variable values (positive values) that are thicker than horizons with low values (negative values). If the series had been distorted to produce thick horizons with low values, instead of thick horizons with high values, the resulting spectra would have been identical.

as the second, third and fourth harmonics, etc. (treating the fundamental as the first harmonic – adopted here). Confusingly others call the peaks at $2f$, $3f$, $4f$, etc., the first, second and third harmonics, etc. The relative amplitude of harmonics depends on the degree to which the signal is non-sinusoidal. As the record becomes more non-sinusoidal the amplitude of the harmonics increases at the expense of the fundamental. The maximum relative amplitude (spectral power) of the harmonic peaks is equal to that of the fundamental. When filtering records that have non-sinusoidal oscillations, it is necessary to use multiple pass bands (a comb filter, Section 4.3) in order to recover the correct shape of the dominant oscillations (Fig. 5.8).

Schiffelbein and Dorman (1986) devised a way to modify the depth scale of a deep-sea sediment record so that time–depth distortions are minimized, as judged from the shape of the resulting spectrum. However, they were concerned with measurements that are independent of sediment composition (e.g. oxygen isotopes) so that the observed

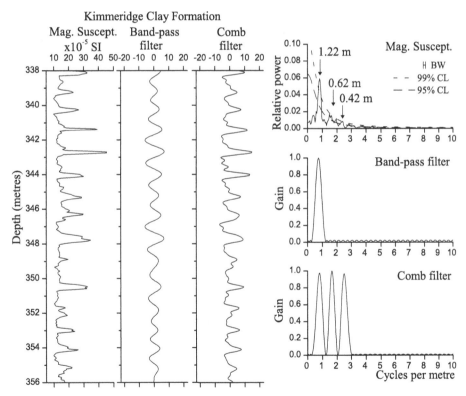

Fig. 5.8 When time-series oscillations are not sinusoidal, filtering can only recover the original oscillations successfully if multiple pass bands are used (i.e. a comb filter, see Fig. 4.4). The data illustrated come from magnetic susceptibility measurements on cores from the late Jurassic Kimmeridge Clay Formation in Dorset, England (Fig. 1.8, Table 1.1, Weedon *et al.*, 1999). The spectral peaks labelled with wavelengths of 0.62 m and 0.42 m correspond, given the uncertainty in frequency denoted by the bandwidth (BW), to the second and third harmonics of the 1.22 m peak (i.e. close to frequencies of 2/1.22 m and 3/1.22 m).

variable is an accurate reflection of the amplitude of the original signal, and thus a simple guide to the distortions of the time–depth relationship. By applying distortions of varying amounts 'in reverse' they modified their records until they were more-or-less sinusoidal so that harmonic and combination tone peaks (discussed below) were largely eliminated from the spectrum.

However, consider time series of $\%CaCO_3$ in deep-sea sediments. If the variation in carbonate flux causes the accumulation rate variations, then the resulting depth-domain record involves two distortions of the carbonate content signal. Firstly, there is the modification of the time–depth relationship – causing harmonics and combination tones. Secondly, the shape of the measured carbonate variations will be changed in a non-linear manner by the variation in carbonate flux since this is a percentage measure – also causing harmonics and combination tones (Ricken, 1991b; Herbert, 1994). Thus, unlike the oxygen isotopes used in the Schiffelbein and Dorman (1986) procedure, $\%CaCO_3$ is not a simple guide to the signal-driven distortions in accumulation rate. Application of their method to a $\%CaCO_3$ record could thus potentially minimize the time–depth distortions, but the carbonate values in the modified time series would not reflect the changing carbonate flux (the signal that supposedly led to the time–depth distortion) or the carbonate contents of the original sediment.

An added complication is that diagenesis will also modify carbonate contents during compaction, for example by carbonate redistribution via pressure dissolution, which results in cuspate or even square-wave carbonate variations (Ricken, 1986; Weedon, 1989; Meyers *et al.*, 2001). In such sections the effects of systematic (as opposed to random) variations in compaction would be identical to signal-driven variations in accumulation rate. Weedon and Read (1995) used a rather crude, but effective, method to remove compaction variations related to lithology in Carboniferous cyclothems. They used standard decompaction formulae appropriate to the average compaction history of each lithology and showed that removal of the compaction had a profound effect on the shape of their spectra (Fig. 5.9).

Conceivably, hiatuses that occur systematically at a particular point during the deposition of sedimentary cycles, such as at the bases of 'simple sequences' observed in sequence stratigraphy, will also result in distortions that are indistinguishable from the effects of signal-driven accumulation rates (e.g. Naish and Kamp, 1997).

Hagelberg *et al.* (1991) showed that the degree to which oxygen-isotopic time-series oscillations were non-sinusoidal could be determined quantitatively using **bispectral analysis**. Bispectral analysis is concerned with the similarity in the non-linear behaviour of pairs of frequencies (e.g. a fundamental and its second harmonic). **Bicoherence** (analogous to coherency) indicates the size of the non-linear relationship (or non-linear correlation) between the two components. **Biphase** (analogous to phase) indicates the phase relationship of two non-linearly related frequency components (King – née Hagelberg –, 1996). If the time series consists of purely sinusoidal oscillations, the bicoherence and biphase will be zero (Priestley, 1988). Time-series oscillations can have narrow troughs and broad peaks (or vice versa) – i.e. a peaked or flattened distribution, rather than a normal distribution of values around the

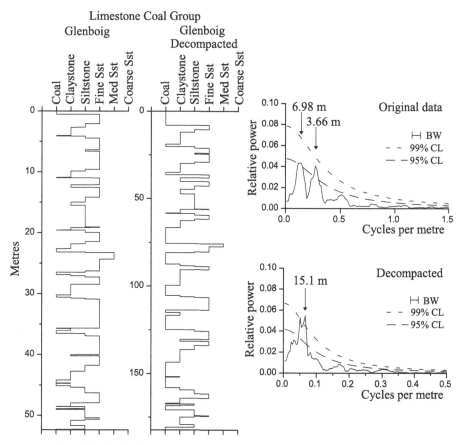

Fig. 5.9 Digitized graphic logs of Carboniferous (Namurian) cyclothems in the Limestone Coal Group at Glenboig (central Scotland, Fig. 1.8, Table 1.1). Using numerical codes from 1 to 6 for coal to coarse sandstone, the spectrum of the data contains two well-defined spectral peaks. Weedon and Read (1995) applied different factors for each lithology in order to remove the distorting effects of compaction. The decompaction factors, derived using appropriate published formulae, were coal = ×16.00, claystone = ×5.071, siltstone = ×2.444, sandstone = ×1.533. The spectrum of the crudely decompacted records contains just one spectral peak. This demonstrates that substantial variations in compaction (or accumulation rates) from lithology to lithology can have a substantial impact on the spectral shape. The methods adopted by Weedon and Read (1995) did not allow for variations in accumulation rates. An alternative procedure for removing the distortions, caused by lithology-dependent variations in accumulation rate and compaction combined, is the gamma method of Kominz and Bond (1990) and Kominz *et al.* (1991, see text). Sst: sandstone.

time-series mean (measured by the degree of kurtosis; Williams, 1984). Alternatively, the oscillations might have a saw-toothed pattern – i.e. a skewed distribution of values around the mean. Provided the bicoherence is significant, the biphase can be used to assess the degree of kurtosis versus the skew of the time series values about the mean. This is because, unlike power spectra and coherency spectra, the phase of the data affects the results in bispectral analysis (King, 1996).

Kominz and Bond (1990) and Kominz *et al.* (1991) introduced **gamma analysis** in order to correct time-depth distortions related to accumulation rate variations within cycles. Gamma is just cycle period divided by cycle thickness. The idea is that in a cyclic sedimentary sequence there will be many cycles of differing thicknesses composed of various thickness combinations of different lithologies. The method assumes that the cycle period is almost constant and that each lithology had a constant and distinct accumulation rate (and compaction factor). There is a natural constraint that no lithology can have an inferred negative accumulation rate – if one does the method fails. Unfortunately, it is not always the case that there is only one dominant regular cyclicity present, or that the accumulation rates are constant for each lithology, so the procedure does not always yield meaningful results. Kominz (1996) used the gamma method on late Pleistocene grey-scale data from a deep-sea sediment core that had previously been orbitally tuned. She showed that with even quite modest accumulation rate variations the method fails or results in rates that produce an incorrect time scale. Hinnov and Park (1998) also found that the gamma method produced results that differed from expectation.

When there is more than one major frequency component present in a time series, non-linear distortions can produce **combination tones** in addition to harmonics (Fig. 5.10). Combination tones are also called **interference beats** and **interference tones**. Like harmonics, combination tones occur at distinct frequencies in relation to the fundamentals. Their generation is known as **intermodulation** or **frequency mixing** (King, 1996). With two fundamentals, at frequencies f_1 (lower frequency) and f_2 (higher frequency), the primary combination tones occur at $f_2 + f_1$ (a **summation tone**) and at $f_2 - f_1$ (a **difference tone**). However, secondary combination tones occur between the fundamentals and harmonics of either fundamental (e.g. at $2f_1 + f_1$, $2f_1 + f_2$, $2f_1 - f_2$, etc). With three or more fundamentals the number of harmonics and combination tones can be very large (Schiffelbein and Dorman, 1986). When a component frequency is higher than a fundamental (i.e. either a harmonic or summation tone) it is referred to as an **overtone**. Difference tones, found at lower frequencies than the fundamental, are called **undertones**. Bispectral analysis can be very useful for examining records with combination tones and harmonics (King, 1996).

Like harmonics, combination tone spectral peaks are enlarged, by time–depth distortions, at the expense of the fundamentals. Which combination tones will be expressed in the spectrum as sizeable peaks is difficult to predict. For example, in the case of a mild non-linear distortion of a series consisting of fundamentals with wavelengths of 2.50 m and 0.70 m (Fig. 5.10), the large amplitude components consist of one harmonic (1/1.25 m = 2/2.50 m) plus three combination tones (1/0.97 m = 1/0.70 m − 1/2.50 m, 1/0.55 m = 1/0.70 m + 1/2.50 m, 1/0.45 m = 1/0.70 m + 1/1.25 m). However, with greater distortion this changes (Fig. 5.10) to two harmonics (1/1.25 m = 2/2.50 m, and 1/0.83 m = 3/2.50 m) plus two combination tones (1/1.59 m = 1/0.70 m + 1/1.25 m, and 1/0.97 m = 1/0.70 m − 1/2.50 m).

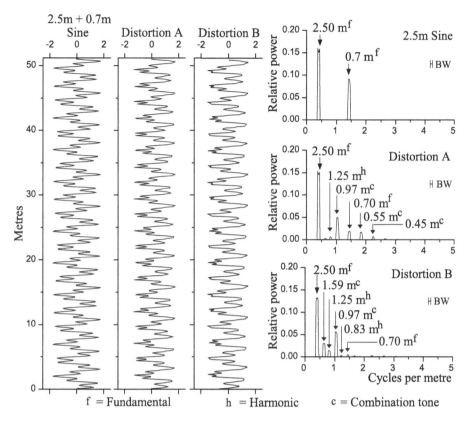

Fig. 5.10 The distortions used in Fig. 5.7 but applied to a time series with two cycles with wavelengths of 2.5 and 0.7 m. The non-linear distortion of the oscillations generates harmonic peaks and, due to the presence of two fundamentals, combination tone peaks as well.

Note that in the second case the higher frequency fundamental is almost invisible using a linear power plot. This means that combination tones or harmonics detected in cyclostratigraphic records could be mistaken for fundamentals, making the use of wavelength ratios to identify the environmental origin of the cyclicity somewhat problematic. However, when the same distortions are applied to the test series a different outcome is observed (Fig. 5.11). After the first distortion, the higher frequency fundamental is just distinguishable from the background noise, but after the more pronounced distortion it is swamped by the noise. This result reflects the greater difference in wavelength and amplitude of the fundamentals when compared to the example in Fig. 5.10. When there is a big difference in wavelength of the fundamentals, the shorter wavelength components are made to vary substantially in wavelength. Thus in this case the high-frequency (1/0.75 m) cycles are stretched, or lengthened, in the maxima of the long wavelength cycles, but squeezed, or shortened, in the minima of the long wavelength cycles. The result is akin to the effects of random variations in accumulation rate; the higher frequency power is spread out over

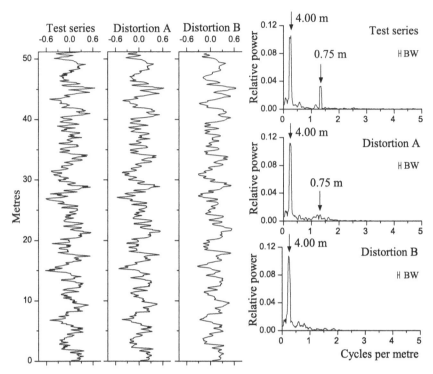

Fig. 5.11 The same distortions as Fig. 5.7 applied to the test series. The large variations in accumulation rate, related to the long wavelength cycles, cause the broadening of the peak associated with the short wavelength oscillations (cf. Fig. 5.5). As a result in this case small harmonic peaks and combination tone peaks cannot be distinguished from the background noise.

a relatively wide frequency band so that only one fundamental can be readily identified (Herbert, 1994).

As well as distortions of signals by signal-driven variations in accumulation rates, imposed amplitude modulation (Section 1.3) causes distortion of sinusoids that always generates both sum and difference tone peaks (Fig. 1.5). In such cases the size (amplitude) of the amplitude modulation, compared to the average amplitude of the main frequency component, determines the size of the combination tone peaks; the larger the imposed amplitude modulation, the larger the tone peaks.

5.3   Cyclostratigraphic signal distortions related to other processes

5.3.1   Rectification

Rectification, first mentioned in the context of continuous-signal records (Section 2.2.1), involves a system that has a non-linear amplitude response to an input

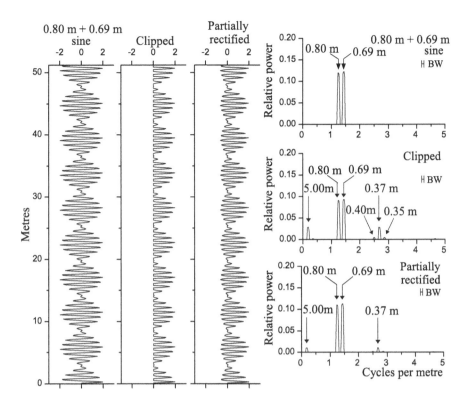

Fig. 5.12 Rectification results from an asymmetric, or non-linear, response of a system to a signal showing amplitude modulation. In this example, a signal resulting from heterodyne amplitude modulation is used. Half-wave rectification results from clipping the data at the mean level. Partial rectification, more likely in cyclostratigraphic data, leads to less pronounced combination tone peaks and harmonic peaks than occur with half-wave rectification.

signal. This means that a signal with varying amplitude, or amplitude modulation, is transformed into a record where the amplitude variation is not symmetrical about the mean (Figs. 2.3 and 5.12). Rectification is caused by non-linear processes in the environment and during sedimentation/diagenesis (e.g. Olsen, 1986; Crowley *et al.*, 1992; Hagelberg *et al.*, 1994). It has been suggested that in some cases the 100,000-year (100 ka) eccentricity cycle in Mesozoic strata was encoded as partially rectified 20-ka precession cycles (Ripepe and Fischer, 1991; Short *et al.*, 1991; Crowley *et al.*, 1992; Herbert, 1992).

A simple form of rectification involves **clipping** or simply resetting all values above, or below, a threshold to the threshold value. For example, **half-wave rectification** involves clipping the data by resetting all the values below the mean, or above the mean, to the mean value (Fig. 5.12). In normal amplitude-modulated data there is no spectral peak at the frequency of the modulation (Section 1.3). Rectification has the effect of producing a time series where the local mean varies in relation to the wavelength or period of the modulation. As a result spectral peaks appear at the modulation frequencies. However, the sharp edges in the clipped data also produce harmonics and summation tones at higher frequencies (Section 5.2.4, Hagelberg *et al.*, 1994). For example, Fig. 5.12 illustrates heterodyne amplitude modulation, using cycles with periods of 1/0.80 m and 1/0.69 m. The rectification produces power at the modulation or beat frequency, which for heterodyne AM occurs at the difference tone (1/5.0 m $=$ 1/0.69 m $-$ 1/0.80 m). Additionally, there is a peak related to a summation tone (1/0.37 m $=$ 1/0.69 m $+$ 1/0.80 m), and two harmonics (i.e. 1/0.40 m $=$ 2/0.80 and 1/0.35 m $=$ 2/0.69 m). Less severe or partial rectification produces smaller combination tone and harmonic peaks (Fig. 5.12).

5.3.2   Bioturbation

Bioturbation affects the properties of the great majority of time series obtained from sediments. Burrowing organisms tend to mix sediment from different stratigraphic levels in the topmost interval. This 'mixing layer' is between a few centimetres and a few tens of centimetres thick. The mixing has an effect like a moving average; high-frequency variations in composition are reduced in amplitude so effectively bioturbation acts as a low-pass filter (Section 4.3, Dalfes *et al.*, 1984). This means that power spectra of bioturbated records are steeper or 'reddened' compared to non-burrowed or laminated records (Fig. 5.13).

Pisias (1983) found that, for pelagic sediments, different net accumulation rates had no significant effect on the slope of spectra for oxygen isotope records. He speculated that the thickness of the mixing zone was related to the accumulation rate. Trauth *et al.* (1997) showed that mixing-zone thickness is correlated with the flux of organic matter to the sea floor. In other words, the depth of burrowing depends on the availability of food. A weak relationship between the depth of burrowing and deep-sea sediment-accumulation rate can be explained in terms of the covariation of productivity, sediment flux and organic matter flux (Trauth *et al.*, 1997). It cannot be assumed that the nature of bioturbation, and its impact, remains constant through time at a particular site (Schiffelbein, 1984). Trauth (1998) provided a mixing model that can be used to simulate the impact of time-varying bioturbation on palaeoceanographic variables.

As for many types of filtering (Section 4.3), bioturbation causes a phase shift in the signal. Bard (2001) showed that bioturbational phase shifts are of negligible importance for records of orbital cycles (with periods of tens of thousands of years and longer) at the accumulation rates of pelagic sediments. Similarly he demonstrated that the smoothing effect of bioturbation is essentially independent of accumulation rate at these time scales (consistent with Pisias, 1983). However, at shorter time scales (less than ten thousand years) phase shifts can be significant and the degree of smoothing is linked to accumulation rates. Additionally, the phase shift affects variables derived from the coarse fraction (e.g. oxygen isotopes from foraminifera)

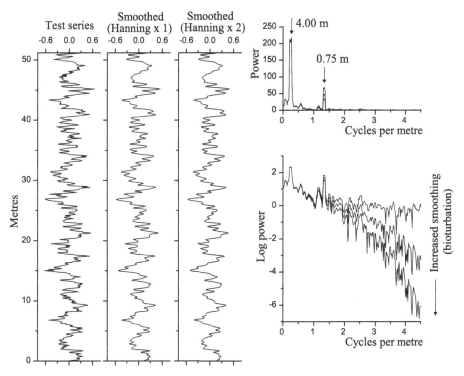

Fig. 5.13 Smoothing of the test series using a three-point weighted moving average (Hanning weights). Increased smoothing leads to increased steepening of the spectrum – analogous to the effect of bioturbation.

differently to those derived from the fine fraction (e.g. oxygen isotopes from nannofossils, Bard, 2001). Anderson (2001) discussed the effects of bioturbation and the influence of accumulation rates on marine sediments recording millennial-scale variability (Section 6.8).

5.3.3   Undetected hiatuses

Major stratigraphic gaps are usually identified and avoided during the generation of cyclostratigraphic records. This reflects the need to produce a record with a thickness scale that has at least an approximately linear relationship with time (Section 2.3.3). Identifiable gaps are indicated by major changes in the character of the record (e.g. at sequence boundaries in continental shelf sediments) and/or sedimentological and palaeontological observations indicating erosion/non-deposition. In some circumstances, such as the analysis of layer-thickness records of tidal cycles, the character of the environmental variable might be known well enough for hiatuses to be detectable, though they can considerably complicate time-series analyses (Archer and Johnson, 1997). Sometimes it is necessary to use separately those data collected from above and below the known gaps. When there is a change in mean accumulation rates across a hiatus, there is a multiplication of spectral peaks for records obtained in the depth domain (Section 5.2.3).

Unfortunately, in the fine-grained sediments frequently used to produce cyclostratigraphic time series, detecting minor stratigraphic gaps via sedimentological observations is exceedingly difficult. Biostratigraphy only helps detect gaps when the gaps are so large that expected biostratigraphic events are missing or amalgamated. If the gaps are sufficiently widely spaced, then wavelet analysis (Section 4.6) is particularly good at hiatus detection in cyclostratigraphic time series (Prokoph and Barthelmes, 1996; Prokoph and Agterberg, 1999).

However, empirical data demonstrate that, on average, estimated rates of accumulation decrease when the time span being considered increases (e.g. Sadler, 1981). This is true for all sedimentary environments investigated, including the pelagic realm (Sadler, 1981; Anders et al., 1987; cf. Ramsay et al., 1994a, 1994b; Spencer-Cervato, 1998). The observation is biased by the over-estimation of rates from currently active sedimentary systems, compaction, etc. (Sadler and Strauss, 1990). Nevertheless, it is widely accepted that the relationship between mean accumulation rates and time span mainly reflects the presence of pervasive, undetected stratigraphic gaps or hiatuses.

Completeness of stratigraphic sections can only be defined meaningfully in terms of some scale of observation (Sadler and Strauss, 1990). For example, a section representing a million years is quite likely to contain some sediment from each 100-ka interval. If it did it would be complete at the 100-ka scale. However, the same section is much less likely to contain sediment from every year. In other words, the completeness at the scale of a year is much smaller than that at the scale of 100,000 years. Theoretically, completeness can be estimated by comparing accumulation rates at one time scale with accumulation rates at another. In practice though, there are many problems with this approach whether applied to real sections or to averages (Sadler and Strauss, 1990). In general, deep-sea sections are more complete than continental shelf and terrestrial sections (Sadler, 1981; Anders et al., 1987).

Nevertheless, Algeo (1993) suggested a method for estimating completeness for long stratigraphic sections with many palaeomagnetic reversal events. Similarly, Schwarzacher (1998) showed that for tuned or un-tuned orbital-climatic cycles (Section 6.9.3) in individual stratigraphic sections, one can estimate completeness. All that are required are rather precise absolute dates from two stratigraphic levels and a count of the number of intervening sedimentary cycles with known orbital periods. Determine the duration implied by the number of preserved orbital cycles, then divide the result by the time span indicated by the absolute dates. This provides an estimate of the completeness at the time scale of the orbital cycles. Clearly, in this situation, one would need a method for deducing the period of the orbital cycles without reference to the absolute age control (e.g. via wavelength ratios or using the characteristic amplitude modulation of the orbital-precession cycle).

A method for estimating relative completeness would be to compare data sets from multiple sites (provided that it is certain that the same time interval is represented and the same environmental signal was recorded stratigraphically at each site).

Thus all stratigraphic sections have undetected hiatuses at some scale. The issues for time-series analysis are: (a) at what scale do the hiatuses occur and (b) what are the effects on the characteristics of the observed cyclostratigraphic record? In terms of the first issue, if the hiatuses are rather small, then it is possible that the time series is complete at the scale of the cycles of interest. Additionally, the hiatuses might be formed as part of the normal cycle development, particularly in the case of sea-level oscillations (e.g. Naish and Kamp, 1997). An example of a complete series might be an orbitally tuned palaeoceanographic deep-sea sediment record from the last 35 million years. The tuning procedure should have revealed whether there is a record from each orbital cycle.

Unfortunately for older, untuned records from the deep sea, or indeed any record lacking an absolute time scale, it is usually impossible to be sure that all the original regular cycles are present. This partly reflects the lack of sufficient precise absolute ages in records from, for example, tidal laminites, fossil growth bands, pre-Neogene sediments, etc. Hence, in many situations one has to accept that there may be undetected hiatuses at the scale of interest. This leads to consideration of the second issue; how might hiatuses affect the time series?

The most widely used model for the generation of hiatuses, perhaps most applicable to siliciclastic continental shelf settings, is based on a random walk model of sedimentation (e.g. Schwarzacher, 1975, 1998; Tipper, 1983; cf. Plotnick, 1986). This is akin to the Brownian walk introduced in Section 3.3.1. In a simple version of this model a random number series (Gaussian white noise) is used to determine whether sediment accumulates, or is removed, from a sediment pile (cf. Strauss and Sadler, 1989). For example, sediment could be added when the random number is positive and removed when it is negative. Since successive random numbers can be negative, 'amalgamated' hiatuses will be formed. Brownian walks are non-stationary (i.e. there are long-term positive or negative trends) so sediment can build up despite the equal probability of erosion or deposition (Turcotte, 1997).

A simplistic version of this model has been used for Fig. 5.14. A 10,000-point run of Gaussian white noise was generated with unit variance (using the algorithms *ran1* and *gasdev* from Press *et al.*, 1992). The random numbers determined the number of sediment increments 0.4 m long added to the new record if positive, or number of sediment increments removed from the new record if negative. Ten thousand values of the target time series (4 m sine + 0.75 m sine + AR(1), sample interval 0.1 m) were used as the stratigraphic record into which the hiatuses were introduced. Two runs are illustrated, each having the same number of values (512), as the unaltered, or complete, series which has been used for comparison. The first run

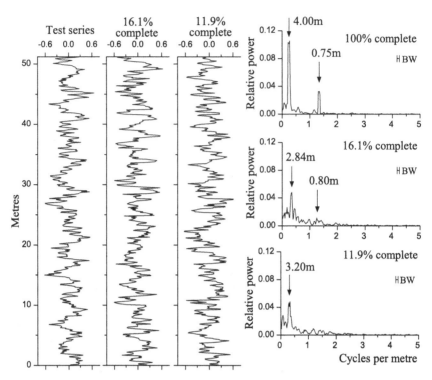

Fig. 5.14 Introduction of multiple hiatuses at random spacing and random amounts of strata missing as applied to a very long test series. The results of two independent runs are shown. In both cases the low-frequency peak has shifted to a higher frequency. In the first case the peak associated with short wavelength oscillations has shifted to a lower frequency. The wavelength ratio has changed in the first case from 5.33 (4.00 m/ 0.75 m) to 3.55 m (2.84 m/0.80 m). The effect of hiatuses on wavelength ratios for depth records requires more investigation.

produced a record that was 16.1% complete at the 0.4 m scale. In other words, a section of the original time series that was 318.8 m long was reduced to 51.2 m. The second run, based on a different run of white noise, was 11.9% complete at the 0.4 m scale.

The model could have used shorter sediment increments with each step. Furthermore, instead of using the random numbers to determine how many sediment increments were added or removed, one sediment increment could have been added or removed each time. Nevertheless, the model, used here for convenience, does capture realistically the type of hiatus distribution expected of a random walk. Figure 5.14 shows that, in both runs, the hiatuses have the effect of adding considerable noise to the spectra. This reflects the addition of power related to spurious oscillations that were produced by the truncation of the regular cycles. The hiatuses mean that the higher frequency spectral peak observed for the original series is either considerably less easily identified or entirely indistinguishable from the background noise after the introduction of the hiatuses. This effect is akin to that discussed for random variations in accumulation rate (Section 5.2.2).

Another effect is that the frequency of the peaks associated with the regular cyclicity has been altered. This apparent frequency shifting of the spectral peaks due to hiatuses, for records from the depth domain, was noted by Weedon (1989). The change in frequency of the major peaks apparently results from dominant shortening or lengthening of the cycles by the abundant gaps in the generated series. In other words, the effects on final cycle lengths has not involved equal lengthening and shortening of the cycles – unlike the case of random variations in accumulation rate. The implication is that in cyclostratigraphic records based on a thickness scale rather than a time scale, the wavelength ratios of pairs of regular cycles may have been altered.

Thus undetected hiatuses at the scale of the cycles of interest, though clearly undesirable, should always be considered. In sections where a reliable time scale is not available (e.g. from multiple radiometrically dated horizons, or orbital tuning), the following should be borne in mind. Firstly, undetected hiatuses may mean that some regular cycles are missing so caution should be used when estimating cycle periods from sparse absolute dating. Normally only a maximum possible period can be inferred. Secondly, hiatuses introduce noise into the spectrum making regular cycle detection harder, particularly at the higher frequencies. Finally, hiatuses can apparently alter the wavelength ratios of any regular cyclicity identified (though this requires further investigation).

5.4   Practical time-series analysis

5.4.1   How long should a time series be?

Whenever a new time-series analysis is planned, one of the issues to arise concerns the amount of data, or number of points, required. Logistical limits to the amount of data collected include the length of accessible stratigraphic section or sediment core and the cost, in time and money, of generating the observations. For a given thickness of strata, the number of points generated depends on the sampling rate. As discussed in Chapter 2 (Section 2.4.2) the primary control on the sampling rate should be the avoidance of aliasing. As well as conducting a pilot study to investigate the smallest scale of significant variability (i.e. variation that could be aliased), it is wise to consider the possibility that the oscillations present are not sinusoidal. In particular, the sampling needs to be dense enough to ensure that the harmonic components of cuspate or square-wave oscillations (Section 5.2.4) are correctly sampled (Herbert, 1994). Note that, in a single stratigraphic section, the oscillations for one variable may be sinusoidal, but for another they may be non-sinusoidal (e.g. Fig. 5.15).

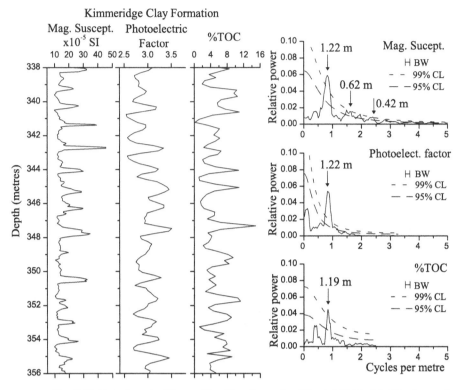

Fig. 5.15 Data from the Kimmeridge Clay Formation (Fig. 1.8, Table 1.1, Weedon *et al.*, 1999; Gallois, 2000; Morgans-Bell *et al.*, 2001). Different time series variables from the same interval of the Swanworth Quarry borehole have different-shaped oscillations. Unlike total organic carbon (TOC) and photoelectric factor, the magnetic susceptibility record produces spectral peaks that indicate the second and third harmonics of the 1.22-m cycles due to the non-sinusoidal oscillations (cf. Fig. 5.8).

While generating time series values it is wise to bear in mind the effects, on power spectra, of interpolation. In the lower two-thirds of the Belemnite Marls carbonate data it has already been demonstrated that aliasing is not a problem when the sampling rate is increased from 3 to 6 cm (Figs. 2.10 and 2.11). Thus in Fig. 5.16 the effect of interpolation of the original data at 6-cm intervals has been compared with data that were decimated to the same sampling interval. The decimated data have a spectrum that is very similar to the original spectrum, though the Nyquist frequency is different of course. In particular the slopes of the spectra are the same. However, the interpolation involves inevitable smoothing which, like bioturbation, acts as a low-pass filter. This produces a spectrum that is considerably steeper at high frequencies compared to the spectra of the original and decimated data (Schulz and Stattegger, 1997). Effectively the interpolation adds a moving average component to the spectral background. Not surprisingly this means that it is not always possible to use the robust AR(1) fitting method of Mann and Lees (1996, Section 3.5) for finding the confidence

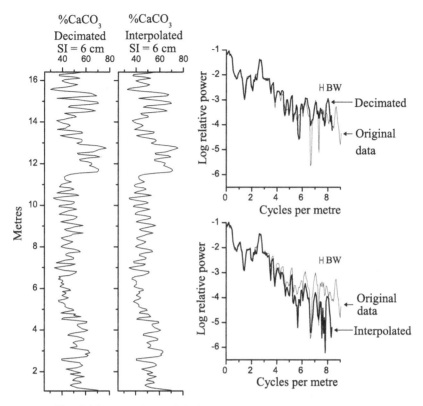

Fig. 5.16 Comparison of decimation and interpolation using part of the Belemnite Marls carbonate data. In Section 2.4 it was demonstrated that there is no significant aliasing of the Belemnite Marls carbonate data when the sample interval is increased from 3 cm to 6 cm. The spectrum of the decimated data closely follows that of the original record. However, interpolation to a sample interval of 6 cm involves smoothing of the data with a consequent increase in the slope of the spectrum compared to the original spectrum. SI: sample interval.

levels of power spectra. Interpolation can be avoided for spectral and cross-spectral analysis by use of an appropriate algorithm (e.g. Press *et al.*, 1992; Schulz and Stattegger, 1997).

Given a sufficiently narrow or short sample interval, how should one judge whether a time series is long enough for meaningful analysis? One issue concerns spectral resolution. If it is suspected that there are two regular cycles present with wavelengths of f_1 and f_2, then in theory they will form two resolved spectral peaks if the data length exceeds $2/(f_1 + f_2)$ (Jenkins and Watts, 1969). In practice though, given factors such as broadening of spectral peaks by random accumulation rate variations, much longer series should be used whenever possible. As well as using a high-resolution spectral estimation procedure such as the maximum entropy or multi-taper method, spectral resolution can be increased by zero-padding (e.g. Muller and Macdonald, 1997a, 1997b). However, zero-padding has the effect of

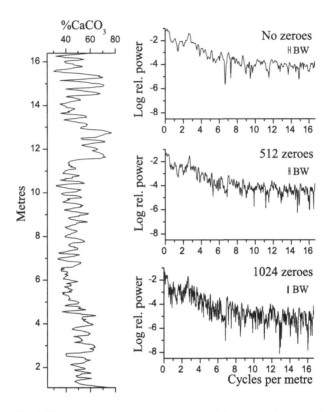

Fig. 5.17 Zero-padding illustrated for part of the Belemnite Marls data. Increased zero-padding, whilst improving the frequency resolution of the spectrum, increases the variance of the spectral estimates and decreases the degrees of freedom. Zero-padding can be used to satisfy a constraint on the number of data points used by a computer algorithm (usually an integer power of two). In such cases the degrees of freedom would be controlled when zero-padding by increasing the smoothing of the spectrum. Alternatively, zero-padding is used to substantially improve the frequency resolution, but this will not necessarily allow narrow peaks to be distinguished from the noisier background (Fig. 3.28).

increasing the variability of the spectral estimates so the degrees of freedom decrease and the confidence levels rise (Figs. 3.28 and 5.17). Alternatively, given zero-padding, increased smoothing of the spectrum, used to maintain a sufficient number of degrees of freedom, would mean that the final spectrum does not have an increased frequency resolution.

In general then, the most satisfactory way to increase spectral resolution is to increase the length of the time series (Section 3.3.2). However, random variations in accumulation rate mean that whereas a low-resolution spectrum might reveal a single spectral peak, a high-resolution spectrum may reveal multiple peaks. This effect is seen even with data based on accurate time scales. For example, in the case of orbital precession, the number of frequency components identified depends on the resolution bandwidth because this is a quasi-periodic, not a periodic, phenomenon (Fig. 5.18).

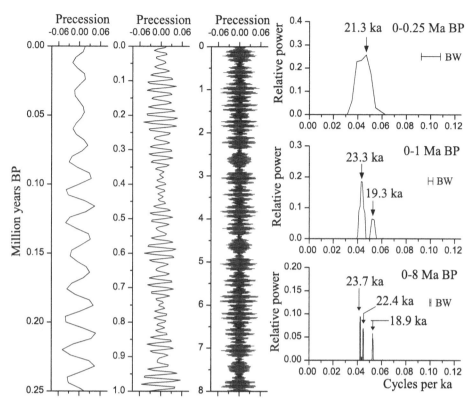

Fig. 5.18 Orbital precession over different time intervals based on the Laskar *et al.* (1993a) (1,1) orbital solution (Section 6.9.2). The number of frequency components or 'regular cycles' associated with orbital precession, as judged using the power spectra on the right, increases with increasing frequency resolution.

The way that the appearance of the spectrum of precession variations changes with record length raises the issue of what constitutes a 'regular cycle'. Thus far in this book it has been the identification of spectral peaks that has been used as the criterion needed to detect the presence of regular cycles. However, if the presence of a single spectral peak defines 'regularity', then precession oscillations would not be described as 'regular' for most time series used to investigate orbital-climatic forcing. A more logical definition might consider the variation in wavelength of the oscillations. Thus individual precession oscillations mainly vary in period from about 13,000 to 27,000 years or by a factor of about 2. Hence, one could define regularity in terms of the uncertainty in the wavelength/period (or frequency) associated with a spectral peak. For example, the resolution bandwidth (Section 3.3.2) can be used to measure the uncertainty in the wavelength or period of oscillations giving rise to statistically significant spectral peaks. When the uncertainty implies that the length of the oscillations varies by a factor of two or less, one could declare that regular cyclicity had been detected. Of course there is no reason why a factor less than two should not be used, and a smaller uncertainty is certainly desirable.

Box 5.1. How long does a time series need to be?

We need to determine N, the number of time series points required in a record with a constant sample interval. Suppose that the longest regular cycle we are interested in has a frequency f and wavelength λ $(= 1/f)$. As discussed in the text we can use the criterion that the uncertainty in the frequency of this regular cycle must be equal to or less than a factor of two. Clearly a more stringent factor could be used if desired. The resolution bandwidth (BW) is used to define the uncertainty in frequency (i.e. $\pm^1/_2$BW). Thus we have:

$$2 = \frac{f + {}^1/_2\text{BW}}{f - {}^1/_2\text{BW}}$$

rearranging this gives:

$$f = \frac{3\text{BW}}{2}$$

or

$$\lambda = \frac{2}{3\text{BW}} \tag{1}$$

For most spectral estimates (i.e. excluding the multi-taper and maximum entropy methods) the bandwidth can be defined (Bloomfield, 1976) as:

$$\text{BW} = \frac{1}{V \times N \times \text{SI}} \tag{2}$$

where V is the variance of the spectral window and SI is the sample interval. Using equations (1) and (2) we have:

$$\lambda = \frac{2 \times V \times N \times \text{SI}}{3} \tag{3}$$

It is common in cyclostratigraphic studies to use eight degrees of freedom (or more). Since the degrees of freedom (DOF) are defined (Bloomfield, 1976) using:

$$\text{DOF} = \frac{2}{V}$$

with eight degrees of freedom we have $V = 0.25$. Inserting this value into equation (3) yields:

$$\lambda = \frac{N \times \text{SI}}{6}$$

so that:

$$N = \frac{6\lambda}{\text{SI}}$$

This means that, in this case, the time series will only be 'long enough' when there are at least six repetitions of the longest wavelength or period of interest. Naturally the frequency uncertainty criterion and the degrees of freedom can be modified as necessary.

It is straightforward to establish how long a time series should be given the regularity criterion based on the frequency uncertainty of the cycle of interest (see Box 5.1). Thus, knowing the approximate wavelength, or period, of suspected regular cycles in a stratigraphic record, in general one would need a time series that was at least six times longer. Crucially, the presence of a statistically significant spectral peak is not sufficient to be able to claim the detection of *regular* cyclicity. An estimate of the uncertainty in the wavelength or period is required as well. This is especially important in the lowest frequency part of a spectrum. It is unfortunately the case that some authors have claimed the detection of regular cycles when they have less than six examples in their record. This situation arises partly because of the plotting conventions for spectra in cyclostratigraphic studies. Using power spectral plots with power or log power versus frequency, one observes peaks at low frequencies. However, use of log power versus log frequency renders spectral maxima, where the uncertainty in wavelength or period is rather large, as very broad features that are much less likely to be mistaken for indications of regular cyclicity (e.g. Figs. 3.11 and 3.31, Wunsch, 2000a).

Another consideration is the reliability of the spectral peaks, or the stationarity of the time series. The simplest check for stationarity is to split the record in two, and then check for the presence of significant peaks at the same frequencies in each part. This then implies that given a suspected regular cycle with a particular wavelength or period, one should, if possible, generate a time series that is at least 12 times as long.

5.4.2   Interpreting spectral peaks

The methods used to distinguish spectral peaks from the spectral background have already been discussed (Section 3.5). However, it is worth considering several practical issues.

Different variables observed in the same stratigraphic section may contain evidence for different regular cycles. For example, at Ocean Drilling Program (ODP) site 722 southeast of Oman inorganic geochemical ratios acting as proxies for siliciclastic grain size and hence wind speed (Ti/Al and Cr/Al) show predominant 23-ka variations related to the intensity of the Indian Monsoon (Fig. 5.19, Weedon & Shimmield, 1991). The monsoon drives seasonal upwelling and thus delivery of nutrients from intermediate waters to the surface of the Arabian Sea. However, proxies for nutrient levels (Ba/Al and P/Al), although varying at 23 ka, are dominated by 100-ka variations (Fig. 5.19). Thus, it appears that the overall level of nutrients at the surface of the Arabian Sea is determined by global ocean circulation – which is linked to global ice volumes and hence the 100-ka eccentricity cycle in the late Pleistocene (Weedon and Shimmield, 1991).

Since the recognition of spectral peaks uses a statistical procedure one can never be certain that a particular peak is the product of some sort of regular environmental variation rather than just an example of large-amplitude noise. All that can be inferred is that a peak has a certain likelihood of representing noise. For example, if a spectral peak exceeds the 90% confidence level, there is a one in ten chance that it represents noise.

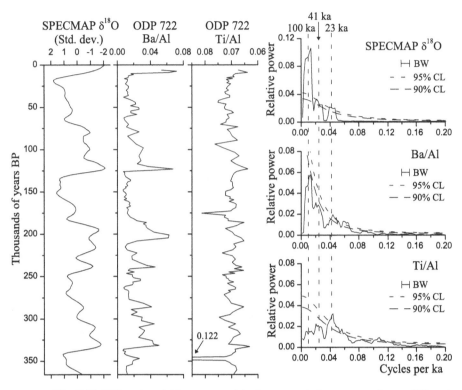

Fig. 5.19 Different variables measured in the same strata may reveal very different modes of variability. The data illustrated were obtained from ODP site 722 on the Owen Ridge southeast of Oman (see Fig. 1.8, Table 1.1, Weedon and Shimmield, 1991). At this site Ba/Al is a proxy for productivity and mainly varies at a period of 100 ka. Ba/Al varies in a manner that is similar to the SPECMAP stack of oxygen isotopes – which is usually treated as a proxy for global ice volume (Section 6.9.3). On the other hand, Ti/Al, related to siliciclastic grain size and thus wind speed, varies mainly at a period of 23 ka (Weedon and Shimmield, 1991). Note reversed scale for Ti/Al.

Furthermore, one can expect that ten per cent of all the spectral estimates will exceed this level. If more than ten per cent of the estimates in a particular spectrum exceed this level, then there is a strong likelihood that some of them have not arisen by chance. Even in this situation though, it is not possible to decide which spectral peaks are environmentally significant.

How can we increase the confidence that a particular spectral peak represents a 'real' environmental cycle? Firstly, if the peak exceeds a higher confidence level, on the face of it, it is more likely to be real. However, note that if the data have been processed in particular ways then the statistical significance may not be as meaningful as if the data are unaltered. For example, pre-whitening, filtering and orbital-tuning have the effect of concentrating the variance into particular frequency bands. In other words, the distribution of power in the spectral background no longer conforms to that in the 'unaltered' data.

Usually regular environmental cycles persist throughout the whole length of the stratigraphic record. Thus, as mentioned in the last section, one way to check whether a particular spectral peak is meaningful is to split the data into at least two, non-overlapping, subsections and check for the presence of significant concentration of power at the same frequency in each subsection. Of course the frequency resolution of the spectra from the subsections will be worse than the spectrum for the whole data set. Note that even using a good time scale there are regular cycles that do not persist indefinitely, a well-known example being the development of strong 100-ka oscillations in mid-Pleistocene marine oxygen-isotope records (Section 6.9.4b). Nevertheless, checking for a particular spectral peak in data collected from younger and older strata than the original record could increase confidence (Fig. 5.20). Similarly,

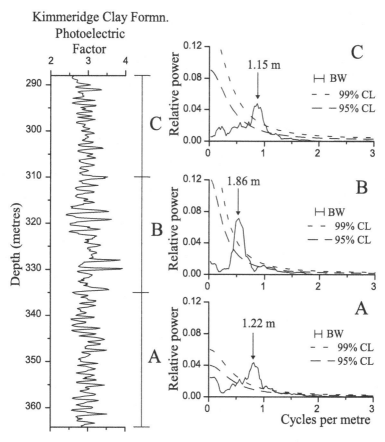

Fig. 5.20 The photoelectric factor log from part of the Swanworth Quarry borehole through the Kimmeridge Clay Formation (Dorset, England, Fig. 1.8, Table 1.1, Weedon *et al.*, 1999; Gallois, 2000; Morgans-Bell *et al.*, 2001). Power spectra for subsets of this downhole log reveal a single well-defined spectral peak. The simplest interpretation is that there were 'long-term' variations in accumulation rate, and that the same regular environmental cycle is represented by sedimentary cycles that varied in wavelength from 1.15 m to 1.86 m.

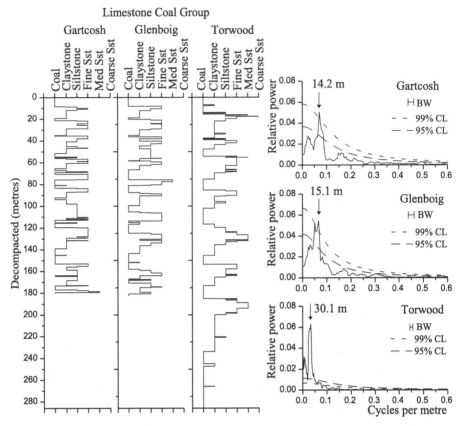

Fig. 5.21 Logs of decompacted mid Carboniferous cyclothems from the Limestone Coal Group of central Scotland (Fig. 1.8, Table 1.1). Figure 5.9 explains the decompaction procedure used. These logs were obtained from precisely the same stratigraphic interval (from the Knightswood Gas Coal to the Hartley Coal) in each location and correspond to the same time interval (Weedon and Read, 1995). However, moving from Gartcosh to Glenboig to Torwood the facies change from predominantly deltaic to predominantly fluvial. The predominantly fluvial locations have relatively thick, relatively coarse, channelized sandstones associated with stratigraphic gaps, rather than the thinner, finer, more readily correlated, sheet sandstones encountered in the more deltaic facies. These factors help explain the differences in the power spectra (see Weedon and Read, 1995 for more detailed explanations). Sst: sandstone.

looking for the same regular cyclicity in records collected from adjacent sites will decrease the chance that an incorrect assessment has been made (cf. Fig. 5.21).

As discussed in Section 5.2.4, non-sinusoidal oscillations generate harmonics and combination tones. This means that even if a spectral peak can be considered distinct from the spectral background, this does not automatically mean that it correctly identifies an environmental cycle. Of course harmonics are normally distinguishable from fundamentals because of their lesser power and integer frequency ratios. Unfortunately there are cases where the frequency ratio of the expected cyclicity matches that of a harmonic, for example with orbital-obliquity and precession cycles (Section 6.9). In this situation the only recourse is to inspect the data in the time or depth domain to look for large amplitude variations with the frequency of the peak suspected of being a harmonic. In the obliquity and precession example, one would look for oscillations at the 20-ka scale as opposed to merely cuspate (wide trough and narrow peaks or vice versa) 40-ka oscillations.

Finally, it is worth mentioning the issue of amplitude modulation. As explained in Section 1.3, heterodyne amplitude modulation arises from the summation of two regular oscillations with very similar frequencies (e.g. Fig. 1.4). In orbital-climatic studies in particular the amplitude modulation of precession has often been treated as diagnostic (Fig. 5.18). In other words, environmental cycles, like orbital precession, that are quasi-periodic are associated with multiple, closely spaced, spectral peaks. Similarly, imposed amplitude modulation arises when the amplitude of a high-frequency oscillation is determined by a much lower frequency variation (Fig. 1.5). Imposed AM produces characteristic small-amplitude side-lobes (at the sum and difference tones). Thus it is important when trying to extract regular cyclicity from time series via band-pass filtering (Section 4.3) to allow in the design of the filter for the presence of multiple spectral peaks and side-lobes related to amplitude modulation, otherwise misleading results will be obtained.

5.5   Chapter overview

- Signals of environmental change, as recorded stratigraphically, are subject to distortions caused by factors such as:
  - (a)   variations in accumulation rate,
  - (b)   components of environmental systems exhibiting non-linear responses to the primary signal (e.g. a non-linear response to climate change leading to rectification),
  - (c)   bioturbation,
  - (d)   undetected hiatuses, and
  - (e)   variations in compaction.
- These distortions can significantly affect the shape of power spectra and hence the interpretation of the recorded signal. In particular, the possibility that statistically significant spectral peaks result from harmonics or combination tones needs to be considered.
- In order for a statistically significant spectral peak to be regarded as evidence of *regular* cyclicity, the resolution bandwidth, or frequency uncertainty, must not be too large. Accordingly, the period/wavelength that corresponds to the frequency of the spectral peak must be no more than a sixth of the duration/ length of the time series.
- Confidence in the detection of regular cyclicity in a time series can be increased by checking for the same cyclicity at additional stratigraphic intervals and adjacent sites.

Chapter 6

Environmental cycles recorded stratigraphically

6.1 Introduction

There are a variety of environmental processes that have been used to explain regular cycles encountered in stratigraphic records (in the broad sense from growth bands in living organisms and fossils to layered ice, speleothems, sediments and sedimentary rocks). This chapter provides brief introductions to some of the relevant issues for each process and attempts to give some insight into the nature of the stratigraphic records involved.

The periodic and quasi-periodic processes discussed here range through many magnitudes in terms of their characteristic period and they are discussed in the order of increasing duration. However, all these environmental cycles are expressed via changes in climate and/or, in the case of tidal records and some continental shelf orbital-climatic records, via changes in sea-level. The discussions only consider cycles with a reasonably well-documented stratigraphic expression. Thus, although the day/night cycle, caused by Earth's rotation, apparently accounts for microscopic growth increments in corals (Wells, 1963; Cohen *et al.*, 2001), this aspect of coral growth does not appear to have been studied using time-series analysis. Additionally, there are several types of atmospheric and oceanic variability that are only now being investigated and for which convincing stratigraphic records have yet to be found (e.g. decadal variability in the Indian Ocean, e.g. Saji *et al.*, 1999; Webster *et al.*, 1999). Not discussed are processes not primarily driven by climate and/or sea-level. Such processes include, for example, autocyclic processes in fluvial systems (Schwarzacher, 1993; cf. Peper and Cloetingh, 1995), fault activity (Cisne, 1986; Ito *et al.*, 1999; Morley *et al.*, 2000) and volcanic impacts on global temperatures (e.g. Stuiver *et al.*, 1995; Briffa *et al.*, 1998a).

In most examples discussed it is controversial both whether and how the environmental cyclicity produced stratigraphic records (e.g. Burroughs, 1992). Note that in many cases the assessment that a particular environmental cycle accounts for an observed regular stratigraphic cycle has been based on no more than the presence of a power-spectral peak with the 'correct' frequency. Ideally the chain of processes linking environmental changes to the stratigraphic signature should be sought. However, it is rarely possible, in the case of older records, to be sure exactly how a particular stratigraphic signal was generated, often because there is no way to choose between plausible alternatives (Weedon, 1993). Instead the combination of repeated, independent identification of the same forcing via rigorous time-series analysis based on accurate chronologies, together with the lateral and temporal persistence of a signal, have been sufficient to persuade doubters that the proposed explanation is worth considering. In view of the many factors that can lead to signal distortion (Chapter 5), it is inevitable that not all aspects of a particular stratigraphic record will be a perfect match for the implied forcing. Therefore, although time-series analysis provides procedures for quantitative and qualitative descriptions of stratigraphic records, the interpretation of the results should always be considered to be open to question.

In this chapter, the spectra illustrated with the exemplary data sets have been generated using the multi-taper method (Section 3.4.3) with the number of tapers varied according to the amount of zero-padding, such that there were eight degrees of freedom in each case. Apart from Section 6.2, confidence levels were located using the robust AR(1) method of Mann and Lees (1996, Section 3.5). The locations of the cyclostratigraphic data sets illustrated are shown in Fig. 6.1, and further information is provided in the figure captions and Table 1.1.

6.2   The climatic spectrum

Before considering individual processes that account for the cyclicity observed in the field of cyclostratigraphy, it is worth considering the nature of the climatic spectrum. Mitchell (1976) discussed meteorological and climatic phenomena with periods ranging from hours to the age of the Earth. He inferred the nature of the spectrum that would result from a perfect, long record of temperature. The theoretical climatic spectrum he illustrated consisted of narrow and broad peaks, related to particular phenomena, superimposed on a weak red noise background.

Shackleton and Imbrie (1990) attempted to reconstruct the climatic spectrum using multiple deep-water oxygen isotope records from deep-sea sediments as proxies for climatic variations (global ice volume). Their records had time scales ranging between 1000 years and 130 million years. They confirmed earlier observations that, contrary to the Mitchell (1976) concept, the climatic spectrum at periods greater than a thousand years has a rather pronounced red noise character with a log variance density versus log frequency slope of -1.5. This is reminiscent of the spectral slope of -2.0 expected of a record of a Brownian or random walk, where successive

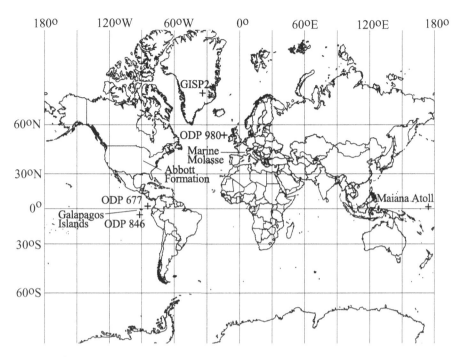

Fig. 6.1 Location map for the cyclostratigraphic records illustrated in Chapter 6. Table 1.1 provides an outline description of these data sets.

observations just represent the accumulation of white noise variations (Section 3.3.1). Another possibility is that the climate represents a chaotic system with a large number of interacting factors (dimensions) producing self-organized criticality and a history with a power law spectrum (Section 4.7, Dessai and Walter, 2000).

However, it now appears that the climatic spectrum does not exhibit a simple power law. Pelletier (1997) effectively repeated the procedure of Shackleton and Imbrie (1990), but used a large number of direct temperature plus oxygen-isotope records from ice cores with time scales ranging from one day to 200,000 years. This revealed, for marine locations, a mean (log power versus log frequency) climatic spectrum with a slope of -0.5 for periods of one day to 2000 years, a slope of -2.0 for periods of 2000 years to 40,000 years and a flat spectrum for periods of 40,000 to 200,000 years. This sequence of power-law 'behaviours' over different frequency ranges was accounted for using a model of convective heat transport in the atmosphere.

The climatic spectrum has been re-created here following methods similar to Shackleton and Imbrie (1990) and Pelletier (1997). A series of spectra have been combined to produce a spectrum spanning periods from one year to a million years. The individual spectra were obtained from the well-documented oxygen-isotope records from the Greenland ice core projects (GRIP: see http://www.esf.org/esf_article.php?language=0&article=166&domain=3&activity=1, and GISP2: see http://www.gisp2.sr.unh.edu/GISP2, results discussed by Hammer *et al.* (1997), the data sets were obtained from these web sites and from the Greenland Summit Ice Cores CD-ROM).

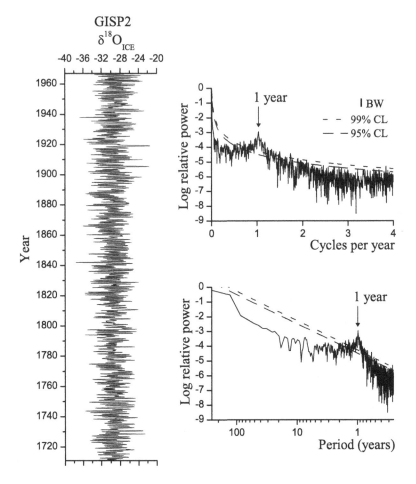

Fig. 6.2 $\delta^{18}O$ obtained at a sample interval of 0.125 years from the GISP2 ice core for the years 1711.25–1967 AD (Fig. 6.1, Table 1.1, see Section 6.2 for the source of these data). The associated spectrum reveals very strong evidence for the annual cycle (Section 6.4).

For longer period variations, oxygen-isotope data were obtained from benthic foraminifera from east Pacific site V19-30 plus Ocean Drilling Program sites 677 and 846 (data from http://delphi.esc.cam.ac.uk/coredata/v677846.html, Shackleton *et al.*, 1990, 1995a, 1995b).

These various data sets are illustrated in Figs. 6.2 to 6.6 and involve sample intervals ranging from 0.125 years to 3000 years. Note that in each case the isotopic records start from the present day, so that each record provides different time perspectives extending to a few hundred years ago up to a few million years ago. The advantage of combining spectra obtained from records with different sample intervals is that a very large range in frequency can be investigated with relatively small data sets. For example, to obtain a record sampled at 0.125 years extending to 5.5 million years would require 44 million measurements. The records used here, although they

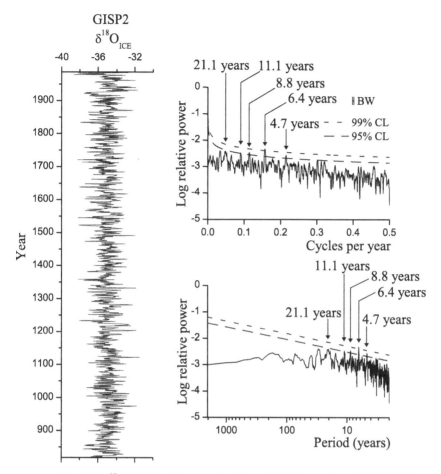

Fig. 6.3 $\delta^{18}O$ obtained at a sample interval of 1.0 year from the GISP2 ice core for the years 818–1987 AD (see Section 6.2 for the source of these data). The associated spectrum reveals evidence for cycles with periods of 4.7, 6.4 and 8.8 years that may be associated with the North Atlantic Oscillation (Section 6.6). There is also weaker evidence for cycles with periods of 11.1 and 21.1 years that may be related to solar variability and sunspot numbers (Section 6.7).

are unusually long for cyclostratigraphic investigations, amount to just 5977 data points.

The longest available direct temperature record is from central England. Mean temperature was determined for each season (i.e. a sample interval of 0.25 years) and runs from 1659 to 1996 (Fig. 6.7, Plaut *et al.*, 1995, data from Hulme and Barrow, 1997). These data were used for comparing, at short periods, a directly measured temperature spectrum with the spectra from the ice core records.

In this chapter section, confidence levels were obtained using power law models for the log power versus log frequency spectra (Section 3.5). The spectra were combined by simply truncating the low frequency end of one spectrum (where the log-frequency resolution is low) at the Nyquist frequency of the next spectrum (where the

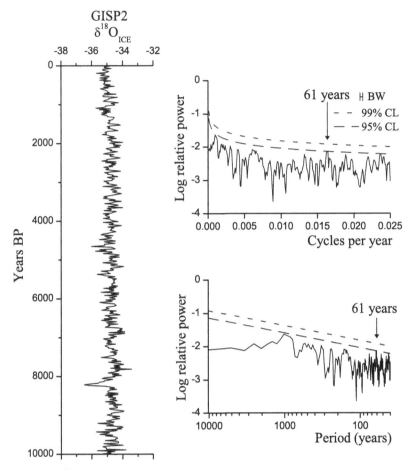

Fig. 6.4 $\delta^{18}O$ obtained at a sample interval of 20 years from the GISP2 ice core for 0–10,000 years before present (see Section 6.2 for the source of these data). The associated spectrum reveals evidence for a cycle with a period of 61 years that might be related to the North Atlantic Oscillation (Section 6.6). BP: before present.

log-frequency resolution is high). Despite the use of the same proxy in every case, there were offsets of the backgrounds of the spectra partly due to the use of different intervals for channel sampling the ice cores. To correct for this the spectra were simply offset in terms of log power so that a continuous spectral background extended smoothly from the lowest to the highest frequencies. The result was a combined set of oxygen isotope spectra with varying log-frequency resolution. To produce a combined spectrum with a consistent log-frequency resolution, the spectra were interpolated, thereby smoothing the high-frequency parts of the individual spectra (Fig. 6.8).

This method for estimating a climatic spectrum inevitably has limitations. For example, the variations in oxygen isotopes in oceanic bottom water are linked to both temperature and global ice volumes (Shackleton and Opdyke, 1973; Shackleton, 2000). On the other hand, oxygen isotopes in Greenland ice are controlled primarily by

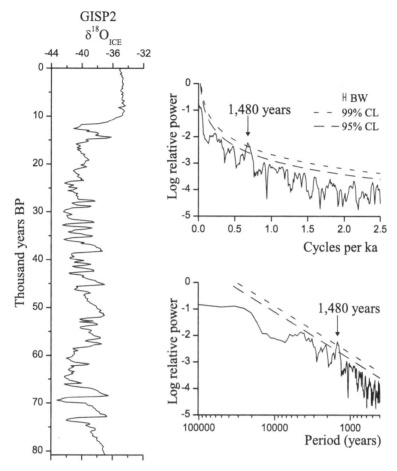

Fig. 6.5 $\delta^{18}O$ obtained at a sample interval of 200 years from the GISP2 ice core for 0–80,800 years before present (see Section 6.2 for the source of these data). The associated spectrum reveals strong evidence for the millennial-scale cycles with a period of about 1500 years (Section 6.8). BP: before present.

atmospheric temperatures (e.g. White *et al.*, 1997); The isotopic data are inevitably smoothed because of sample homogenization, bioturbation in the deep-sea sediments and ice-crystal boundary diffusion (Johnsen *et al.*, 1997; Rempel *et al.*, 2001). Furthermore, for some of these data sets, further smoothing has occurred because of the data interpolation needed to produce records uniformly sampled in time. This variable smoothing must have increased the slopes of the various spectra (Sections 5.3.2 and 5.4.1). It probably partly accounts for the offsets in the spectral backgrounds of the various records, but it is difficult to assess how much the spectral slopes have been changed.

Another problem concerns stationarity. Inspection of the ice core record sampled at 20 years reveals relatively constant $\delta^{18}O$ back to 10 ka BP, where the record was truncated (Fig. 6.4). However, the millennial-scale variability characterizing the last

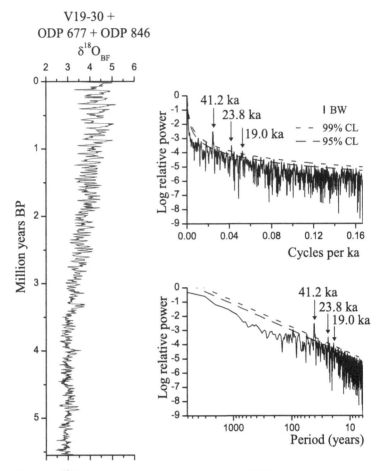

Fig. 6.6 $\delta^{18}O$ interpolated at a sample interval of 3000 years from deep-sea sediment cores V19-30, ODP 677 and ODP 846 in the eastern equatorial Pacific (Fig. 6.1, Table 1.1, data from http://delphi.esc.cam.ac.uk/coredata/v677846.html, described by Shackleton *et al.*, 1990, 1995a, 1995b). The associated spectrum reveals very strong evidence for the Milankovitch orbital cycles (Section 6.9) with periods of 41.2, 23.8, and 19.0 ka (Section 6.9). $\delta^{18}O_{BF}$: $\delta^{18}O$ from benthic foraminifera (i.e. deep-water isotopes).

80 ka (Fig. 6.5) is absent over the Holocene in the Greenland ice core record. Furthermore, it is well known that over the late Pleistocene the response to Milankovitch orbital forcing has varied, particularly at the 100-ka period (Fig. 6.6, Fig. 6.31, Section 6.9.4b), e.g. Pisias and Moore, 1981; Ruddiman *et al.*, 1989). Thus these records provide an imperfect indication of the variability over different time scales for different time intervals back to 5.55 Ma.

Despite the problems with combining spectra based on different sample intervals, the reconstructed climatic spectrum in Fig. 6.8 conforms to the results of Shackleton and Imbrie (1990) and Pelletier (1997) rather well. In particular, the (log-log) spectral background has a slope of around −0.5 for periods of 1 year to about 800 years, from

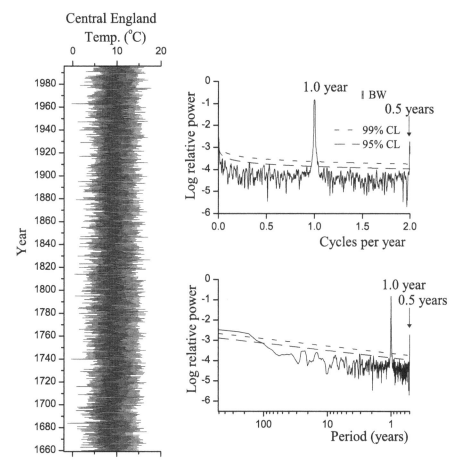

Fig. 6.7 Central England temperatures at intervals of 0.25 years from 1659.25 to 1995.75 AD (data from Hulme and Barrow, 1997). The associated spectrum reveals exceptionally strong evidence for the annual cycle and its second harmonic (Section 5.2.4).

800 to 40,000 years the slope is around −2.0, between 40,000 and 200,000 years the slope drops to about −0.5 again and for periods of 200,000 to one million years the slope resumes at about −2.0. At periods of one year to about 100 years the Central England temperature record produces a spectrum with a very similar appearance to the isotope spectra. The interpolation of the individual isotope spectra has inevitably obscured narrow, but small, spectral peaks. However, the dominant modes of climatic variability over the last 5.5 million years, which have periods of 1 year, around 1500 years and 41,000 years, are readily distinguished as spectral peaks. Despite the caveat concerning the 'steepening' of the isotope records, it should now be apparent that the spectra obtained from any cyclostratigraphic records involving a climatic origin must be tested against a red noise and not a white noise null model (Section 3.5).

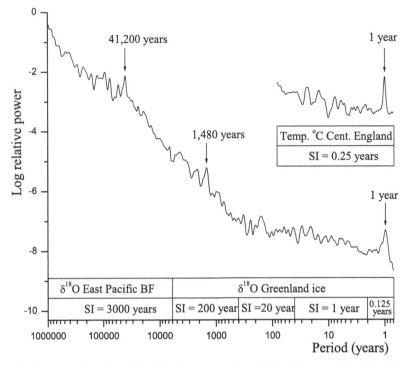

Fig. 6.8 Approximation of the climatic spectrum for periods of between 0.7 and 1 million years based on interpolation of the spectra from isotopic records in Figs. 6.2 to 6.6 and the directly measured temperature record in Fig. 6.7. The Central England temperature spectrum shows the same general pattern as the isotopic records over periods of 0.8 to 100 years. BF: benthic foraminifera; SI: sample interval.

6.3   Tidal cycles

6.3.1   Tidal cycles and orbital dynamics

Over the last two decades, there has been a proliferation of studies of sediments exhibiting stratification related to tidal control. This reflects the improved ability of sedimentologists to recognize tidally controlled deposition and the importance of the resulting information concerning past orbital characteristics. To explain the large number of orbital parameters involved in tidal studies, simplified diagrams have been used here. For greater detail and more precise explanations refer to Pugh (1987). The discussion presented here is based on equilibrium tidal theory, but real tidal systems are complicated by local physical factors such as basin geometry, and the interaction of basin geometry with tidal waves arriving from oceans (e.g. Sztanó and de Boer, 1995).

In order to understand the origin of tidal forces, start by considering the Earth as a solid body, ignoring its rotation (e.g. Lowrie, 1997). Points on the Earth are orbiting the Earth–Moon centre of mass. The centre of mass is located 4671 km from the Earth's centre in the direction of the Moon (i.e. within the planet, Pugh, 1987). Each

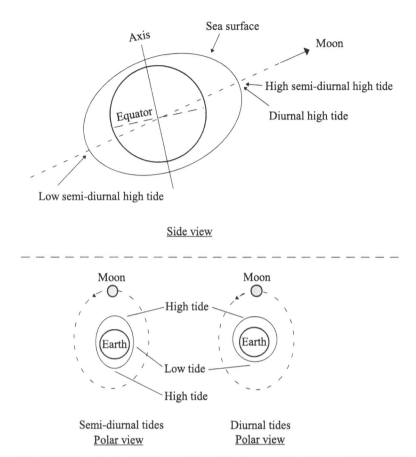

Fig. 6.9 Tides raised by the Moon and caused by the centrifugal force of the Earth–Moon system and the local gravitational forces. Not to scale.

point in the Earth is orbiting at the same rate, producing a common centrifugal force that is directed *away* from the Moon. By analogy consider an ice-skating parent and child holding hands and spinning round each other. If the child lets go both skaters move away from the centre of mass due to the centrifugal forces. Points on the Earth adjacent to the Moon experience a gravitational attraction that is slightly greater than the centrifugal force so a tidal bulge is raised (p. 38 of Lowrie, 1997). On the other side of the planet the gravitational attraction due to the Moon is slightly less than the centrifugal force so a counterpart tidal bulge is formed (Fig. 6.9).

The rotation of the Earth on its axis means that most points on the Earth's surface will encounter two tidal highs and two lows during one day or a so-called semi-diurnal tidal system (Fig. 6.9). However, some places have a single tidal cycle each day or a diurnal system. Other locations sometimes have a semi-diurnal pattern and at other times a diurnal pattern – this is known as a mixed tidal system. The most significant controls on tidal elevations concern a variety of orbital parameters as explained below (Pugh, 1987). However, synoptic weather systems, storms and prevalent winds also influence tidal heights.

Semidiurnal tides

1) Conjunction (New Moon): spring tides   2) Quadrature (Waxing Moon): neap tides
   Perigee: high spring tides

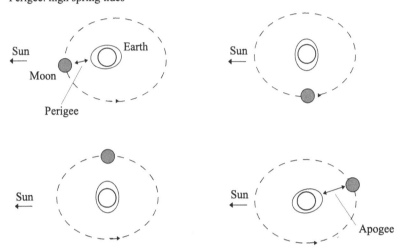

4) Quadrature (Waning Moon): neap tides   3) Opposition (Full Moon): spring tides
                                              Apogee: low spring tides

Fig. 6.10 The fortnightly (two weekly) neap-spring tidal cycle is linked in semi-diurnal systems to the relative positions of the Moon and Sun. Large tidal ranges (spring tides) occur just after alignment of the Earth, Sun and Moon. Not to scale.

The most important variations in tidal heights relate to the neap-spring cycle. For semi-diurnal systems, large or 'spring' tidal ranges occur every two weeks a few days after the alignment of the Moon with the sun (called syzygy). The alignment of these three planetary bodies occurs at both conjunction when the Moon is 'New' and during opposition at Full Moon (Fig. 6.10). At other times the tidal range is smaller (i.e. 'neap' tides) than for spring tides because the arrangement of the Earth–Moon–Sun system is such that the tide raised due to the Sun is not coincident with that due to the Moon. The tidal range is at a minimum at neap tides just after quadrature (waxing half Moon or waning half Moon). From conjunction to conjunction currently takes 29.5 days – the period of the lunar or synodic month. Hence semi-diurnal spring tides occur every 14.8 days.

In diurnal systems spring tides are related to the inclination of the Moon's orbit, rather than alignment of the Moon and Sun. The Moon's orbit is inclined at 5° (varying from 4° 58' to 5° 19', Pugh, 1987) to the plane of Earth's orbit (also known as the ecliptic). The inclination of the Moon's orbit means that it moves north and south of Earth's equator (Fig. 6.11). The angle of the Moon relative to the Earth's equator is known as the lunar declination. In the nodical or tropical month it takes 27.2 days for the Moon to oscillate through a complete declination cycle. This means that in diurnal systems spring tides occur at maximum declination every 13.6 days, with neap tides occurring when the Moon passes the Earth's equatorial plane.

<u>Diurnal tides</u>

1) Moon north of equator: spring tides
   Perigee: high spring tides

2) Moon crossing equatorial plane
   Neap tides

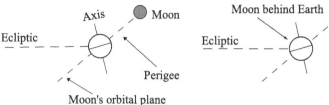

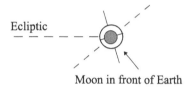

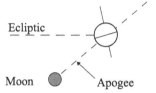

4) Moon crossing equatorial plane
   Neap tides

3) Moon south of equator: spring tides
   Apogee: low spring tides

Fig. 6.11 In diurnal systems the neap-spring tidal cycle is related to the apparent elevation (declination) of the Moon relative to the equator. The declination varies as the Moon moves in its orbit north and south of the equator. Not to scale.

The nodical month tidal cycle adds a small diurnal component to semi-diurnal tides, thus producing an alternation of smaller and larger high tides. This is referred to as the diurnal inequality. The size of the inequality varies with the lunar declination. The effect is to produce a larger first high tide each day, until the Moon passes the equatorial plane, when there is no diurnal inequality, then the second high tide of the day becomes the larger as the declination increases (de Boer *et al.*, 1989). This change in which of the two tides is larger produces a 'cross-over cycle'. Diurnal inequality in semi-diurnal and mixed tidal regimes can be magnified in the higher parts of the intertidal zone (e.g. Archer and Johnson, 1997).

The size of the spring tide is affected by the distance of the Moon. Since the Moon, like all planetary bodies, moves in an ellipse, its distance from the Earth varies. Maximum spring tides in semi-diurnal regimes occur every 27.6 days (the period of the anomalistic month) during the closest approach of the Moon or perigee (Fig. 6.10). The result is an alternation of high and low spring tidal ranges known as the fortnightly inequality (Fig. 6.12). (For completeness, the time taken for the Moon to orbit the Earth is called the sidereal month which lasts 27.3 days, Pugh, 1987).

These various features of tidal height are illustrated (Fig. 6.12) for hourly intervals throughout January 1997 for the semi-diurnal regime at the River Tees, NE England. The hourly tidal elevations were reconstructed using the computer program

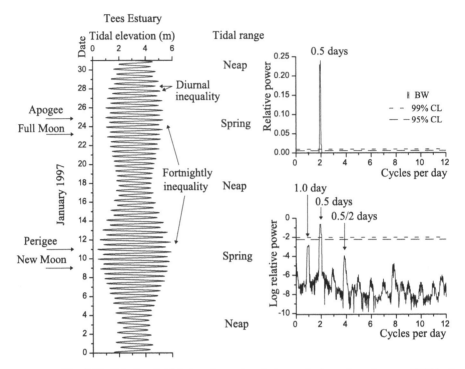

Fig. 6.12 Calculated tidal elevation at hourly intervals throughout January 1997 in the Tees estuary (NE England). This is a semi-diurnal system with a weak diurnal inequality. The spectrum is dominated by the half-daily (semi-diurnal) tidal cycle. However, the oscillations are not perfectly sinusoidal so the log spectrum contains peaks related to harmonics (e.g. 0.5/2 days). The diurnal inequality also produces a small spectral peak with associated higher harmonic peaks.

of Whitcombe (1996) with the tidal constants for the River Tees from the Admiralty Tide Tables (Hydrographic Office, 1996). A continuous record of tidal elevations yields a spectrum containing a single spectral peak corresponding to the period of the basic tidal cycle – i.e. daily or half-daily.

However, stratigraphic records of tidal cycles are not continuous-signal records (Section 2.2.1) of tidal elevation. Instead they are related to the tidal height or amplitude at discrete time intervals, usually once per tidal cycle, as illustrated for high-tide elevations over 1997 in Fig. 6.13. In other words, stratigraphic records of tidal activity are periodic discrete-signal records (Section 2.2.2) where the thicknesses of successive sediment increments are correlated with the tidal range. For example, providing the water speed is not so fast that erosion occurs, the thickness of sand/mud couplets or tidal laminae is often a function of tidal velocity, which is dictated by the tidal range (Visser, 1980; Allen, 1981; Dalrymple and Makino, 1989). Hence thicker sediment increments are produced during larger tidal ranges.

In studies of tidally controlled sedimentation, whether the record relates to a semi-diurnal, diurnal or mixed tidal regime is of prime importance because it provides an indication of the time increments involved (half a day or a day). In many tidally influenced stratigraphic records either the flood- or the ebb-tidal currents are dominant, so generating a single tidal lamina or tidal bundle for each tidal cycle. However, in some settings it is possible for silt or fine sand deposition to occur during the decelerating parts of both the flood- and ebb-phases so creating two laminae per tidal cycle. This would mean that in a semi-diurnal system there would be four laminae deposited each day. Clearly the determination as to whether particular tidal laminae relate to deposition during one or both phases of the tidal cycles is of considerable importance for time-series analysis. Fortunately, deposition during both phases is rare and when it does occur the thickness characteristics of the resulting laminae are sometimes diagnostic (Archer *et al.*, 1995).

The fortnightly neap-spring cycle will often dominate the variation in tidal-laminae or tidal-bundle thicknesses and thus the associated power spectrum (Fig. 6.13). Unfortunately, simply counting the number of increments in neap-spring cycles does not necessarily indicate whether a semi-diurnal, diurnal or mixed system is involved

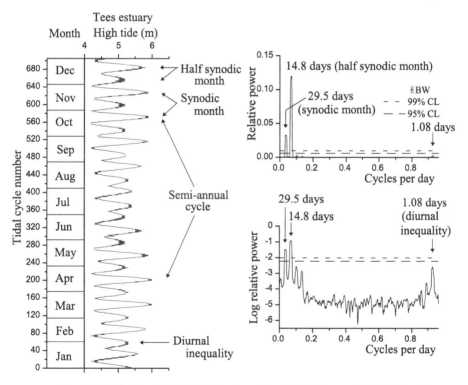

Fig. 6.13 Calculated high-tide elevations (not full tidal range) in the Tees Estuary throughout 1997 (cf. Fig. 6.12). This represents a periodic discrete-signal record of the type represented by stratigraphic records of tidal cycles. There were 705 high tides during the 365.24 days of the year so the sample interval specified during spectral analysis was 0.518 days. The time series shows clear evidence for the fortnightly neap-spring cycle (the half synodic month, 14.8 days) or high spring tides to low spring tides. Additionally, the plot of high tidal elevations reveals the synodic month (high spring tides to high spring tides, 29.5 days) and the semi-annual cycle (highest semi-diurnal spring tides in spring and autumn).

(Archer *et al.*, 1995; Archer and Johnson, 1997). For example, when tidal laminae are deposited on mud flats (instead of as tidal bundles in migrating bed forms), an incomplete stratigraphic record is formed when small high tides fail to reach the site of deposition. An added complication is that in mixed tidal systems some sediment increments represent a day while others represent half a day. Fortunately the presence of a diurnal inequality showing 'cross-over cycles' (Fig. 6.12) is indicative of a semi-diurnal system (de Boer *et al.*, 1989; Archer, 1996; Archer and Johnson, 1997; Kvale *et al.*, 1999). The diurnal inequality produces a spectral peak at or near the Nyquist frequency (i.e. a period of about a day) because successive sediment increments (laminae or silt/mud couplets) are alternately thick and thin (Figs. 6.12 and 6.13).

There are longer period variations in tidal heights. Since the Moon's orbit round the Earth is tilted at about 5° relative to the equator, conjunction or opposition (spring tides) only occur with the Moon *directly* in line with the Sun and Earth twice a year (when the Moon passes over the equator). This produces very large spring tidal ranges in semi-diurnal systems, currently in spring and autumn (Figs. 6.13 and 6.14). In

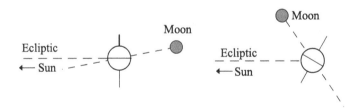

Semi-annual cycle

Northern hemisphere summer
Low semi-diurnal spring tides
High diurnal spring tides

Northern hemisphere autumn
High semi-diurnal spring tides
Low diurnal spring tides

Northern hemisphere spring
High semi-diurnal spring tides
Low diurnal spring tides

Northern hemisphere winter
Low semi-diurnal spring tides
High diurnal spring tides

Fig. 6.14 The semi-annual cycle is related to the alignment of the Sun, Earth and Moon with respect to the Earth's orbital plane (the ecliptic). The figures are drawn looking towards the Earth as though following it around its orbit. In summer and winter the Moon has a large declination relative to the ecliptic causing high spring tides in diurnal systems. In spring and autumn the declination of the Moon to the ecliptic is at a minimum (the three bodies are aligned) leading to high spring tides in semi-diurnal systems. Not to scale.

diurnal systems the tidal range depends on the declination of the Moon as well as the declination of the Earth's orbital plane relative to the Sun or, more simply, the declination of the Moon relative to the ecliptic (Earth's orbital plane). In spring and autumn during the equinoxes, with the Sun over the equator, the declination is at a minimum. Hence, in diurnal systems the largest spring tidal ranges occur in winter and summer (Fig. 6.14). As well as semi-annual cycles, the elliptical nature of the Earth's orbit produces an annual tidal cycle with larger tides during perihelion (closest approach of the Earth to the Sun, currently in the northern hemisphere winter).

Normally the projection of the long axis of the Moon's orbit does not pass through the Sun. However, the long axis rotates in space with a period of 8.85 years. This means that conjunction/opposition coincide with perigee during the March or September equinox every 4.5 years (Pugh, 1987). This produces very high semi-diurnal spring tides during the perigee (or apsides) cycle (Fig. 6.15). Rotation of the Moon's tilted orbit relative to the Earth means that at some times the Moon exhibits a large declination relative to the Earth's orbital plane (current maximum $23.5° + 5° = 28.5°$). At other times the declination relative to the ecliptic is much smaller (current minimum

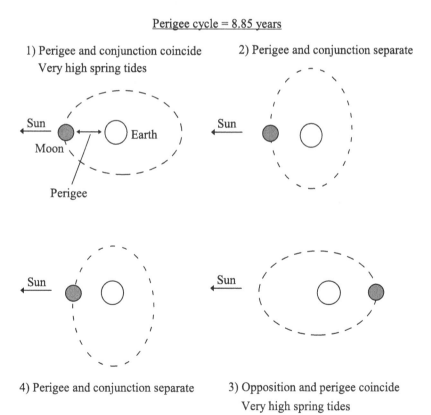

Fig. 6.15 Rotation of the long axis of the Moon's orbit in space causes the perigee cycle. This causes very high spring tides every 4.5 years in semi-diurnal systems. Not to scale.

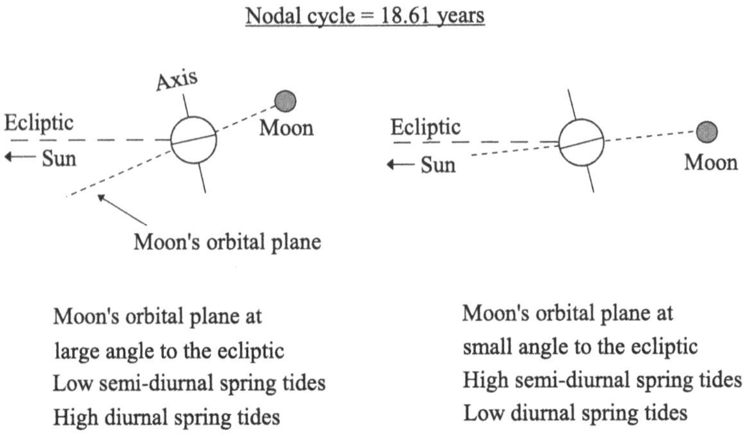

Fig. 6.16 The nodal cycle relates to the changing direction of tilt of the Moon's orbital plane relative to the ecliptic. Not to scale.

$23.5° − 5° = 18.5°$, Fig. 6.16). This so-called nodal cycle, or Saros cycle, lasts 18.61 years (Pugh, 1987; Oost et al., 1993). Finally, every 1600 years perigee and conjunction/opposition coincide with perihelion (Archer, 1996; Keeling and Whorf, 2000; Munk et al., 2002).

Modern tidal predictions are often treated as though a limited number of quasi-periodic components, related to the various orbital components, are sufficient for most purposes. Indeed harmonic analysis is often possible with tidal records (Emery and Thomson, 1997). This gives the impression that stratigraphic records of tidal elevations will have spectra with a few line components and a small white-noise background related to measurement errors. However, Percival and Walden (1993, pp. 468–78), for example, demonstrated a clear red-noise component to a spectrum of tidal height measurements. This they attributed to a series of processes related to weather and local topography. Hence, cyclostratigraphic records related to tidal processes are just as likely to have spectra with a red-noise background as records of climatic processes.

6.3.2   Stratigraphic records of tidal cycles

There is very convincing evidence that tidal signals are encoded in modern sediments (e.g. Visser, 1980; Smith et al., 1990; Oost et al., 1993). As mentioned the stratigraphic information is encoded as periodic discrete-signal records (Section 2.2.2). Ancient strata that are the product of lateral accretion in tidal flows, or vertical accretion on tidal mud flats, or vertical accretion associated with deltas are found from the Archean to the Recent (Fig. 6.17, e.g. Allen, 1981; Allen and Homewood, 1984; Kvale et al., 1989; Smith et al., 1990; Chan et al., 1994; Archer and Johnson, 1997; Eriksson and Simpson, 2000). Thick-thin alternating silt mud couplets and the extremely regular cyclicity are particularly diagnostic of tidal cyclicity (Yang and Nio, 1985; Williams,

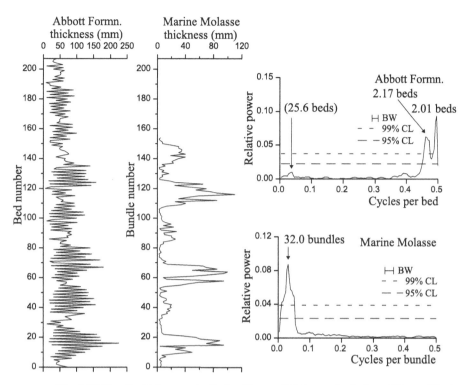

Fig. 6.17 Bundle thicknesses in Carboniferous tidal foreset beds from Abbott, Illinois and Miocene estuarine tidal bundles from Auribeau, France (Fig. 6.1, Table 1.1). These data were obtained with permission from http://www-personal.ksu.edu/~aarcher and are described, along with time-series analyses, by Archer (1996). The data from Abbott reveal a very strong diurnal inequality producing a large spectral peak at the Nyquist frequency (cf. Figs. 6.12 and 6.13). The neap-spring cycle is weakly developed. At Auribeau the diurnal inequality is not apparent, but the neap-spring cycle is very marked and appears to have lasted about 16 days (32 tidal cycles). There is also a very marked fortnightly inequality.

1989), though a good understanding of the sedimentological context of deposition and evidence for cycle periods are essential. For example, the classic Proterozoic Elatina Formation laminites were initially believed to reflect glacial lake varves with implied annual laminae thicknesses modulated by sunspot cycles (Williams, 1981; Williams and Sonett, 1985). Later work from coeval sediments with higher accumulation rates revealed additional characteristics most plausibly explained via tidal control (Williams, 1988, 1989). Note that the time-series analytical descriptions remained valid and useful despite the reassessment of the laminae as diurnal rather than annual, and despite the huge increase in the inferred accumulation rate (Sonett *et al.*, 1988).

Bivalves can also encode tidal information in growth increments (House and Farrow, 1968; Evans, 1972). However, microscopic coral growth increments, as opposed to annual growth bands, appear to be linked to the day/night cycle that is caused, of course, by Earth's rotation (e.g. Wells, 1963; Cohen *et al.*, 2001). Murakoshi *et al.* (1995) described late Pleistocene tidal sediments with semi-diurnal and mixed regime silt/mud couplets demonstrating the same tidal variations as observed in the associated fossil infaunal, intertidal bivalves. The molluscs have aragonite growth increments whose thickness depended on the time the enclosing sediment was submerged. However, the growth lines (organic-rich aragonite layers) have a thickness that depended on the time the sediment was emergent. The bivalves preserve the semi-diurnal inequality, the neap-spring cycle and the fortnightly inequality.

Most studies of diurnal pre-Pleistocene tidal cycles have evidence for the neap-spring cycle and the fortnightly inequality (e.g. Yang and Nio, 1985; Tessier and Gigot, 1989; Chan *et al.*, 1994; Miller and Eriksson, 1997). However, in view of the difficulties encountered with the Elatina record, the presence of the semi-diurnal pattern of thick-thin alternating laminae or tidal bundles is a much more reliable indicator of tidal control (Fig. 6.17, e.g. Kvale *et al.*, 1989; Räsänen *et al.*, 1995; Archer, 1996; Archer and Johnson, 1997). Note that varved (annually laminated) sediments have yielded evidence for the 18.6-year nodal cycle (Schaaf and Thurow, 1997) probably reflecting a long-term tidal influence on climate, also implied by meteorological and ice-core data (Currie, 1987; Stuiver *et al.*, 1995).

Tidally generated laminae or tidal bundles (silt/mud couplets) are not always readily interpreted as the product of tidal factors alone. For example, exceptionally thick tidal bundles have been related to storms (Yang and Nio, 1985; Murakoshi *et al.*, 1995). The position of deposition relative to mean tidal elevation is a critical factor in determining the nature and completeness of the record (Archer and Johnson, 1997). Incomplete semi-diurnal and mixed regime records can be difficult to distinguish from diurnal records. However, using neap-spring cycles, if more than 28 tidal events per semi-lunar month (fortnight) are present diurnal systems can be excluded (leaving either a mixed or semi-diurnal system as a possible explanation). Similarly when more than 50 events are observed per neap-spring cycle a semi-diurnal system can be assumed (Archer and Johnson, 1997).

6.3.3   Records of tidal cycles and the Earth's orbital parameters

Global tidal friction has the effect of reducing the rate of rotation of the Earth (i.e. gradually increasing the length of the day). Currently the rate of rotation is decreasing at the rate of 2.4 milliseconds per century. This results in a transfer of angular momentum to the Moon, which is gradually receding from the Earth (presently at 4 cm per year). In addition to this, in coastal and shelf areas, a considerable fraction, perhaps half, of the world's tidal friction occurs over rough topography on the ocean floor (Egbert and Ray, 2000). This may affect ocean circulation and hence, indirectly, climate (Wunsch, 2000b). Since plate tectonic movements and sea-level changes affect coastlines, shelf area and ocean-floor topography, tidal friction is unlikely to remain constant over geological time (Kvale *et al.*, 1999). The number of neap-spring tidal cycles per year, and the length of the nodal cycle, are particularly useful for determining the length of the day and the Earth–Moon distance in the past (Sonett *et al.*, 1988; Williams, 1989; Archer, 1996; Sonett *et al.*, 1996; Sonett and Chan, 1998; Kvale *et al.*, 1999). However, a detailed understanding of the records of tidal variations allows additional orbital parameters to be estimated (Archer, 1996; Kvale *et al.*, 1999).

The long-term (hundreds of million-year) change in Earth–Moon distance has had an impact on the periods of the orbital-precession and orbital-obliquity cycles (i.e. two of the Milankovitch cycles, Section 6.9, Berger *et al.*, 1989). Recent work on tidally controlled strata and past orbital parameters (Kvale *et al.*, 1999) necessitates modifications to calculations of the past periods of these two orbital cycles. For example, Berger *et al.* (1989) and Berger and Loutre (1994) used estimates of the semi-major axis of the Moon in the past which are considerably shorter than the latest estimates. They also used a model assuming fairly uniform rates of change in the Earth–Moon distance, but the details of the varying rates of change require much more work on stratigraphic records (e.g. Lourens *et al.*, 1996; Kvale *et al.*, 1999; Eriksson and Simpson, 2000; Lourens *et al.*, 2001; Pälike and Shackleton, 2000).

6.4   Annual cycles

The annual cycle, also known as the seasonal cycle, is of course extremely well known. It arises due to the obliquity (or tilt) of the Earth which, in combination with the annual orbit, causes the progression of the seasons related to the direction in which each hemisphere is inclined in relation to the Sun (Fig. 6.18). The annual cycle causes, in most places, very pronounced, essentially periodic variations in climatic variables

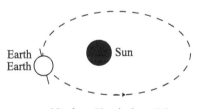

Northern Hemisphere Winter

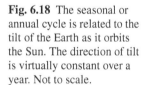

Fig. 6.18 The seasonal or annual cycle is related to the tilt of the Earth as it orbits the Sun. The direction of tilt is virtually constant over a year. Not to scale.

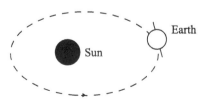

Northern Hemisphere Summer

such as temperature and rainfall (e.g. Figs. 4.9 and 6.7). Changes in obliquity and precession affect the amplitude and shape of the annual cycle (Section 6.9). As explored in Section 2.2.1, Anderson (1986, 1996) argued that the annual cycle and burrowing of varves explain some stratigraphic records of Milankovitch cycles via rectification and smoothing of the annual compositional signal. Kemp (1996) provides a useful collection of papers concerned with annual lamination.

There are many circumstances in which stratigraphic records of annual cycles have been recognized. On land these records are obtained from: annual growth rings in trees (e.g. Briffa *et al.*, 1998b, 1998c; Briffa, 2000); speleothem laminae (e.g. Holmgren *et al.*, 1999; Qin *et al.*, 1999); tufa (Matsuoka *et al.*, 2001); ice-sheet layering and stable isotopes (Fig. 6.2, Alley *et al.*, 1997); and as sediment varves in both high-latitude lakes (Anderson and Dean, 1988; Boygle, 1993; Petterson, 1996) and in low-latitude evaporites (Anderson, 1982, 1984). In marine environments annual cycle records are found as growth bands in corals (Gagan *et al.*, 2000) and in sediments as lamina-scale (less than 1 cm) couplets or multiplets (Bull *et al.*, 2000) forming marine varves (Kemp, 1996). There are of course many ways in which varves are expressed compositionally (Gorsline *et al.*, 1996; O'Brien, 1996). In some settings the lamination may not be particularly distinct making measurements more difficult (O'Brien, 1996; Alley *et al.*, 1997; Pilskaln and Pike, 2001). Lack of bioturbation is essential for the preservation of sedimentary laminations and for varves usually indicates dysoxic or anoxic bottom waters, hyper-salinity or firm substrates (Kemp and Baldauf, 1993; Kemp *et al.*, 1995; Pearce *et al.*, 1995).

Corals provide continuous-signal records (Section 2.2.1) in the form of sub-annually collected stable isotope ratios (Fig. 6.19, e.g. Roulier and Quinn, 1995; Quinn *et al.*, 1998) and geochemical measurements apparently related to sea-surface temperature (e.g. Corrège *et al.*, 2000). The high-resolution measurement of light reflectance in laminated sediments also produces continuous-signal records (Zolitschka, 1996; Schaaf and Thurow, 1997).

However, in many cyclostratigraphic studies of sediments, ice sheets, speleothems and biogenic growth bands, it is assumed that the layering is most plausibly explained by the annual cycle. By treating the laminae as annually produced, periodic discrete-signal records can be generated by measuring layer thicknesses (Section 2.2.2). Care is needed because in marine and lacustrine settings each varve may consist of multiple laminae, such as biogenic laminae related to local productivity and siliciclastic laminae related to runoff (Bull *et al.*, 2000). In such cases the thickness of individual laminae needs to be analysed separately, because the thickness of each varve as a whole does not represent variations in a single environmental variable (Section 2.3.2).

There are cases, such as in the modern Black Sea (e.g. Crusius and Anderson, 1992; Pilskaln and Pike, 2001) and Santa Monica Basin off California, where not all the sediment laminae formed annually (e.g. Gorsline *et al.*, 1996; Hagadorn, 1996). In modern settings sediment-trap studies can be used to demonstrate annual laminae formation (Thunell *et al.*, 1993, 1995). But for older laminated sediments the inference of an annual rather than, for example, a semi-diurnal or diurnal tidal origin is not

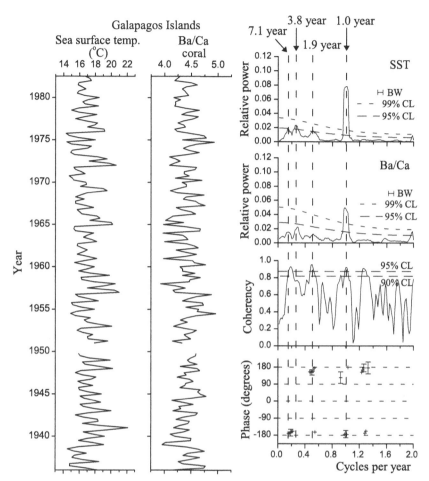

Fig. 6.19 Sea-surface temperature (SST) and a nutrient proxy (Ba/Ca) measured in coral from the Galapagos Islands (eastern equatorial Pacific, Fig. 6.1, Table 1.1, data from Shen *et al.*, 1992). Sea-surface temperature undergoes variations in relation to changing degrees of upwelling of relatively cool, nutrient-rich intermediate water. The power spectra and cross spectra (Section 4.5) demonstrate that most variation occurs over a year with coherent and inversely related changes in SST and nutrient levels (Ba/Ca). Minor changes occur over periods of a few years in relation to the Quasi-biennial Oscillation (1.9 years) and El Niño/Southern Oscillation (7.1 years and 3.8 years, Section 6.5).

straightforward in the absence of absolute dates (Williams and Sonett, 1985; Williams, 1988, 1989; Sonett *et al.*, 1988).

In tree ring studies detection of sub-annual 'false rings' requires comparison of multiple records (Stockton *et al.*, 1985; Schweingruber *et al.*, 1990). Tree-ring thicknesses have been used to infer temperature changes over hundreds of years (e.g. Mann *et al.*, 1998; Esper *et al.*, 2002). However, the interpretation of tree-ring widths, as a straightforward temperature indicator, even after appropriate calibration, is complicated by the effect of increased atmospheric CO_2 on growth rates in the twentieth century, as well as by the need to detrend individual records to remove the effect of tree ageing on ring widths (Briffa *et al.*, 1998b, 1998c; Briffa and Osborn, 1999; Esper *et al.*, 2002).

6.5 The El Niño/Southern Oscillation

6.5.1 The El Niño/Southern Oscillation system

The Southern Oscillation Index (SOI) refers to the changing difference in atmospheric pressure between Tahiti and Darwin, Australia (available at http://www.cgd.ucar.edu/ cas/catalog/climind). The changing pressure difference is linked to changing sea-surface temperature (SST) distributions across the equatorial and tropical Pacific. In the eastern Pacific region, the climatic impacts of disturbances to the normal situation are referred to as El Niño and La Niña. The El Niño/Southern Oscillation (ENSO) system has been reviewed by Philander (1990), Cane (1992), Bigg (1996) and Fedorov and Philander (2000) and useful background papers were edited by Diaz and Markgraf (1992).

Generally movement of warm surface water from the east to the west equatorial Pacific, linked to the trade winds, creates the western Pacific warm pool with sea-surface temperatures above 29°C. The trade winds cause upwelling due to surface divergence, from intermediate depths, of relatively cool, nutrient-rich water to the surface in the eastern equatorial Pacific (Fig. 6.19). Forming part of the Walker circulation, the trade-wind air moving westwards becomes warm and moist and, on rising, causes rainfall in the west. The now dry air returns eastwards high in the atmosphere and then descends over the relatively cool sea surface in the eastern Pacific. The equatorial Pacific forms a coupled ocean-atmosphere system with strong feedbacks. For example, the difference in SST between the east and west, partly caused by the trade winds, helps determine the strength of the trade winds.

At the start of so-called El Niño events, wind bursts blowing eastwards are often associated with the movement of surface water from the western Pacific warm pool towards the eastern Pacific (Webster and Palmer, 1997). The change in the SST gradient leads to the eastwards movement of the main western Pacific rainfall zone – causing drought in Indonesia and Australia. With a smaller SST gradient the trade winds die down or even reverse so that equatorial upwelling in the eastern Pacific generally decreases in intensity or ceases. The decreased upwelling results in warmer east Pacific SSTs and fewer surface nutrients. Additionally, decreased coastal upwelling off Peru and the consequent reduced surface nutrient concentrations lead to the collapse of fisheries off Peru. The now warming, moist and rising air over the eastern Pacific is often linked to heavy rainfall and flooding in western South America, particularly around Christmas ('El Niño' referring to the Christ child).

El Niño events generally last about a year and currently recur approximately every 3–5 years with large events spaced around 3–7 years apart. Sometimes re-establishment of normal SST gradients is quickly followed by development of stronger than normal gradients and an extreme 'normal' climatic regime termed La Niña. La Niña can last

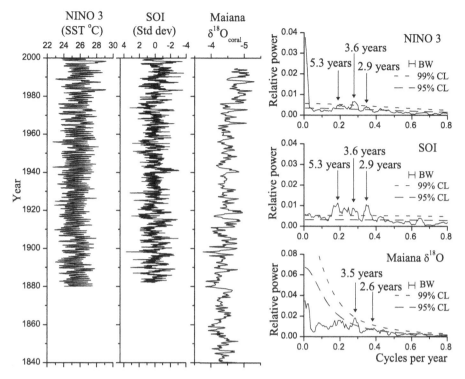

Fig. 6.20 Comparison of sea-surface temperature (SST) in the NINO 3 region, the Southern Oscillation Index (SOI) of pressure differences and δ^{18}O in coral from Maiana Atoll in the west central Pacific (Maiana Atoll, Fig. 6.1, Table 1.1, data from Urban *et al.*, 2000). The associated spectra were estimated using the time interval shared by all three time series. Note that the NINO 3 SST record is dominated by the annual cycle (cf. Fig. 6.19). The Maiana δ^{18}O proxy-record is very similar to the NINO 3 record and the SOI, but has the merit of allowing extension of the record of Pacific climate variability beyond the records of direct measurements (Urban *et al.*, 2000).

for a few years. Because SSTs are strongly linked to the climatic processes in the equatorial Pacific, SST records in particular regions can be used to monitor climatic change as an alternative to the Southern Oscillation Index. For example, Fig. 6.20 illustrates the SST anomaly record from the NINO 3 region defined as 5°S to 5°N and 150°W to 90°W for comparison with the Southern Ocean Index (data available at http://www.cgd.ucar.edu/cas/catalog/climind). The SST record has a much stronger annual component than the SOI, but shows very similar inter-annual variability.

The ultimate cause of the transition from normal to El Niño conditions and back are yet to be understood. The cause is probably linked to the development of slow, eastwards and westwards moving oceanic waves (Kelvin waves and Rossby waves respectively; Bigg, 1996; McPhaden and Yu, 1999). One of the characteristics of ENSO variations is the variability in the recurrence times and amplitudes. One possible explanation for this is that ENSO variability reflects a chaotic system (Section 4.6), due to non-linear resonance of the natural variability of the Pacific Ocean with the annual cycle (Jin *et al.*, 1994; Tziperman *et al.*, 1994).

However, ENSO accounts for the majority of inter-annual variability in the tropical and equatorial Pacific region with important impacts around the world between the tropics and at higher latitudes via atmospheric 'teleconnections' (Philander, 1990). It has long been recognized that ENSO influences the Indian summer monsoon (Bigg, 1996), the annually reversing wind system of the northern Indian Ocean, the Arabian Sea and southern Asia (Barry and Chorley, 1998). In the winter and spring the Asian continent is much cooler than the Indian Ocean, so cooling, sinking air blows seawards. In the late summer the land warms up faster than the sea surface. This produces low pressure and rising air, particularly over the Himalaya Mountains and Tibetan plateau. Replacement of the rising air takes place at low levels from the Indian Ocean. Thus moist air moves landwards, so that the monsoon brings rain to India. In the past El Niño events have caused a disruption to atmospheric circulation in south Asia, raising regional air pressure. This has limited the development of the late summer monsoon. However, the effect of long-term (multi-decadal) rises in mean continental temperatures in Eurasia, perhaps connected to greenhouse warming, has apparently meant that, over the last two decades, El Niño years have not been associated with the absence of monsoon rains (Kumar *et al.*, 1999).

6.5.2   Stratigraphic records of ENSO variability

One of the current areas of research into the ENSO system involves investigation of its variable behaviour (Anderson, 1992a; Quinn, 1992). This has been based partly on the results of computer models and partly on studies of stratigraphic proxy-records. Most time series have been obtained as periodic, discrete-signal records (Section 2.2.2) based on the annual cycle. The proxy-records include those obtained from tree rings (Cleaveland *et al.*, 1992; Meko, 1992), stable isotopes and chemical tracers in corals, as well as organic indices (e.g. Kennedy and Brassell, 1992) and the thicknesses of laminae in sediments. Note that the widely used variations in oxygen isotopes in corals need to be interpreted with great care since some are linked mainly to SST variations (e.g. Shen *et al.*, 1992), some to sea-surface salinity changes (e.g. Tudhope *et al.*, 1995) and some, perhaps most, to a combination of the two (e.g. Quinn *et al.*, 1993).

The Quasi-biennial Oscillation (QBO) represents the approximately biannual reversal in stratospheric wind directions (Labitzke and van Loon, 1990). A link between ENSO and the QBO has been postulated (Lau and Sheu, 1988). Indeed it is often the case that proxy-records used to monitor the history of ENSO variability also involve spectral peaks attributed to the QBO (Fig. 6.19, e.g. Shen *et al.*, 1992; Quinn *et al.*, 1993; Dunbar *et al.*, 1994; Bull *et al.*, 2000; Rittenour *et al.*, 2000). Decadal-scale variations in the ENSO variability have been modelled as resulting from slow movements of parcels of water from high and mid to low latitudes in the Pacific (Latif and Barnett, 1994; Gu and Philander, 1997). There is some evidence that such water movements have occurred (e.g. Zhang *et al.*, 1998; Cole *et al.*, 2000). Dunbar *et al.* (1994) detected 83-, 23- and 11-year cyclicity in a 350-year oxygen-isotope record from a Galapagos Islands coral. As a result they suggested control of sea-surface temperature by solar activity (Section 6.7) which would affect the frequency of ENSO event recurrence.

Over time spans of decades and longer there have been changes in the character of ENSO variability. For example, Torrence and Campo (1998) used wavelet analysis (Section 4.6) to investigate the variable period using the NINO 3 SST records. Urban *et al.* (2000) used an oxygen isotope record from Maiana Atoll in the western central Pacific that is an almost perfect match for the NINO 3.4 SST history back to 1856 (cf. Fig. 6.20). They showed that the modern ENSO event period of about 4 years dates from 1976. Prior to that the period was highly variable back to 1920. At the start of the twentieth century the variability was centred at 2.9 years. In the mid-to-late nineteenth century variability occurred mainly at decadal periods. The Dunbar *et al.* (1994) Galapagos coral oxygen isotope record extends this history back further (cf. Quinn *et al.*, 1998). Eleven-year and 4.6-year period variations characterized the early nineteenth century whilst the 1600s and the 1700s were characterized by oscillations with periods of 3–7 years.

Over longer periods it appears, from a record of siliciclastic laminae deposited in a lake during El-Niño-related storms in Equador, that modern sub-decadal ENSO variability started in the late Holocene about 5000 years ago (Rodbell *et al.* 1999). ENSO-like variability was detected in a SW tropical Pacific coral dating from about 4150 years ago (Corrège *et al.*, 2000). Back to 15,000 years ago, there was apparently no sub-decadal ENSO variability (Rodbell *et al.*, 1999). Between 17,500 and 15,000 years ago, New England proglacial varve thicknesses indicate probable ENSO-related climate variability with periods of 2.5–2.8 and 3.30 years (Rittenour *et al.*, 2000). This record shows that the longer period variability apparently died away between 15,000 and 13,500 years ago, a change attributed to changing orbital-forcing. Bull *et al.* (2000) analysed varves from the Santa Barbara Basin and confirmed the occurrence of ENSO variability during glacial conditions, but in a record dating from 160,000 years ago. Clement *et al.* (1999) modelled the effect of orbital-forcing on the frequency of ENSO events and argued that the recurrence time would have been shorter during warmer episodes in the Pleistocene linked to the orbital-precession cycle. Finally, Ripepe *et al.* (1991) and Fischer and Roberts (1991) argued that the Eocene Green River Formation of Wyoming North America contains varves and that these provide evidence for ENSO variability.

6.6   The North Atlantic Oscillation

The North Atlantic Oscillation (or NAO) is usually defined using the difference in normalized sea-level atmospheric pressure between Stykkishólmur in Iceland and Lisbon, Portugal or Punta Delgada, Azores. Often in winter there is high pressure over the Azores and low pressure near Iceland. Consequently, if an annual index is used, it usually refers to the winter anomaly (December–March). The index is available at http://www.cgd.ucar.edu/cas/catalog/climind. Some authors argue that the NAO should be considered to be part of a more widespread pattern of climatic variability called the Arctic Oscillation.

When the index is high there are frequent strong westerlies transferring heat from the Atlantic to northern Europe in association with cyclones and storms, with high rainfall and relatively high winter temperatures (Hurrell, 1995). Simultaneously southern Europe often experiences low rainfall and low temperatures. When the index is low, blocking anticyclones lead to more northerly winds with cooler temperatures and less precipitation in northern Europe (Fig. 6.21).

Singular spectrum analysis (Section 4.8) indicated that there is a quasi-periodic cycle in temperatures of around 70 years in the North Atlantic region (Schlesinger and Ramankutty, 1994), though this result has since been questioned (Allen and Smith,

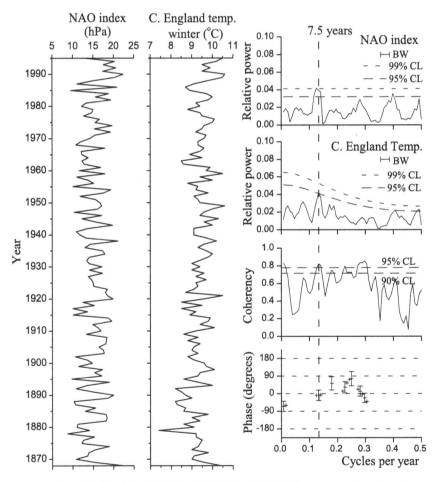

Fig. 6.21 The North Atlantic Oscillation (NAO) index compared to the winter record of central England temperature (NAO and CET data from Hulme and Barrow, 1997). A cycle with a 7.5-year period is coherent and in phase.

1996). However, it should be remembered that the NAO index varies on much shorter time scales, sometimes even between extremes in successive years (Hurrell, 1995, Fig. 6.21). In general most variability is much slower, concentrated at periods of between 13 and 6 years (Appenzeller *et al.*, 1998). Wunsch (1999) argued that the NAO variability is hard to distinguish from red noise. Since the atmosphere is considered to have a relatively short 'memory' of up to a month or so, the similarity between NAO index values in successive winters implies some process having a longer-term memory. The obvious system to provide such a 'memory' is the ocean, but the nature of the feedback between ocean circulation, sea-surface temperatures and atmospheric behaviour over years to decades is controversial (Curry *et al.*, 1998; Rodwell *et al.*, 1999).

Over years to decades, pools of unusually warm or cool water move along the track of the North Atlantic Current towards the British Isles and then swing towards the Labrador Sea (Hansen and Bezdek, 1996; Sutton and Allen, 1997). Before returning southwards the slow-moving pools can, if they are cool enough and dense enough, be mixed downwards to become part of the Atlantic thermohaline circulation (Curry *et al.*, 1998). The sea-surface temperature of these slow-moving pools might partly influence the state of the atmosphere and hence the NAO. The link to the thermohaline circulation could help explain why trade-wind-driven upwelling over the Cariaco Basin off Venezuela, driven by decadal variations in the tropical Atlantic (Chang *et al.*, 1997), is highly correlated with North Atlantic sea-surface temperatures (Black *et al.*, 1999). However, the NAO and associated oceanographic behaviour are unlikely to arise solely from processes within the North Atlantic. For example, although the position of the Gulf Stream shows a 60% correlation with the NAO, much of the rest of the variation in position is correlated with the Southern Oscillation Index of two years earlier (Taylor *et al.*, 1998). Recently modelling has shown a link between the NAO and solar variability (Section 6.7) over decades to centuries (Shindell *et al.*, 2001). Given the importance of predicting weather and climate over decades to centuries, it is likely that considerable effort will be devoted to understanding the full complexity of the NAO over the next few years.

Greenland ice cores first revealed the importance of multi-year and multi-decadal climate variability in the Atlantic in both glacial and interglacial times via measurements of electrical conductivity (Taylor *et al.*, 1993). Subsequently records of stable isotopes (δD and δ^{18}O) from the Greenland Summit cores (correlated with regional temperatures) were also shown to record the NAO (Figs. 6.3 and 6.4, Barlow *et al.*, 1993; White *et al.*, 1997). Tree rings also show variability related to the NAO (D'Arrigo *et al.*, 1993; Cook *et al.*, 1998). Appenzeller *et al.* (1998) attempted reconstruction of the NAO index over the last 350 years using an ice core from West Greenland. They showed, using wavelet analysis (Section 4.6), that the index is highly non-stationary with most variability at less than 15 years, but with 80- to 90-year variability restricted to the last 150 years. A subsequent reconstruction using ice core and tree ring proxies confirmed that the longer period oscillation is not stable through time, but that oscillations with periods of 7–8 and 12.5 years are pronounced between 1750 and 1979 (Cullen *et al.*, 2001). Black *et al.* (1999) studied the concentration of the planktonic foraminifera *Globigerina bulloides* in laminated sediments from the Cariaco Basin from the last 825 years. They argued for 12- to 13-year variations in the NAO indirectly controlling upwelling and surface productivity.

6.7   Solar activity cycles

6.7.1   Sunspot cycles and solar physics

Current investigations using satellites and the SOHO probe are revolutionizing our understanding of solar physics, but there remain many uncertainties over whether certain stratigraphic cycles can be linked to variations in the Sun. Useful reviews of studies of sunspots, solar physics and meteorological/climatic connections are provided by the National Research Council (1994), Burroughs (1992), Hoyt and Schatten (1997) and Beer *et al.* (2000).

Sunspots have been observed for thousands of years. They represent dark patches visible (indirectly) telescopically. The Sunspot Number is routinely used as a proxy for solar variability. The Wolf or Zurich Sunspot Number is defined as ten times the number of sunspots plus the number of sunspots all multiplied by an observer-related constant (p. 36 of Hoyt and Schatten, 1997). The mean number of spots visible each year varies substantially over 8–17 years with a mean period of 11 years (Fig. 6.22). This variation in the mean number of sunspots is called the Schwabe cycle. Berger *et al.* (1990) explore the varying period of these oscillations which according to some are indistinguishable from stochastic or chaotic series (Weiss, 1990).

The sunspots represent areas of cooler photosphere associated with intense localized magnetic fields. Helioseismic tomography has demonstrated that currents converging on sunspot areas down-well/sink to a depth of about 2000 km below the surface of the Sun (Duvall *et al.*, 1996). The currents help to confine the magnetic fields and reduce the rates of heating of the photosphere so that the spots are relatively cool and dark (4200 versus 6000 K). In irregular patches between the sunspots are bright areas named faculae that vary in area and number with the sunspots (Foukal, 1990). Contrary to expectation, satellite measurements of the solar 'constant' showed that over periods of years the total irradiance from the Sun follows the Sunspot Number, at least over two Schwabe cycles (Willson and Hudson, 1988, 1991; Fröhlich and Lean, 1998). Note that the Sun is slightly brighter at sunspot maxima due to more extensive faculae that more than offset the darkening related to sunspots (Foukal, 1990; Hoyt and Schatten, 1997; Fröhlich and Lean, 1998; Rast *et al.*, 1999). It used to appear that there was a weak correlation between neutrino flux and the Sunspot Number, but the correlation has now been shown not to be significant (Hoyt and Schatten, 1997; Walther, 1997). This means that there is no longer a reason to suspect a direct link between solar fusion and sunspots. The eruption hypothesis for sunspot formation held that the variable period of the Schwabe cycle indicated a random generation mechanism at work. However, the phase of sunspot cycles returns to a predictable position if a few cycles are longer or shorter than normal (Dicke, 1978). Generally the length of individual Schwabe cycles

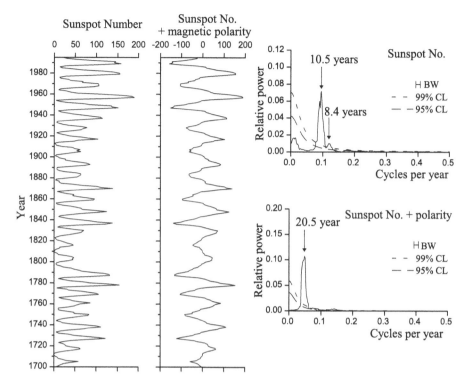

Fig. 6.22 Variations in the Wolf Sunspot Number from 1700 to 1995 (data from Hoyt and Schatten, 1997). The sunspot number is often used as a proxy for solar activity cycles with a dominant period of about 11 years (the Schwabe cycle). The magnetic polarity of sunspots reverses from Schwabe cycle to Schwabe cycle and causes the 21-year Hale cycle. Variation in the amplitude of the Schwabe cycle is termed the Gleissberg cycle (period 70–80 years).

is inversely related to their amplitude. Thus it was suggested that the phenomenon ultimately responsible for creating sunspots occurs periodically at depth, but with a variation in the speed with which the sunspots appear at the surface related to magnetic field intensity (Dicke, 1979, cf. Solanki *et al.*, 2000).

Sunspots are generated in the early part of the 11-year solar cycle singly or in groups at about 35°N and S on the solar surface. Towards the end of the Schwabe cycle they are generated at about 5°N and S. Once formed, the spots and spot groups travel towards the solar equator. At lower latitudes sunspots appear to accelerate across the solar disc as observed from Earth due to the faster motion of the solar photosphere near the solar equator. The photosphere represents the top of an external convection zone. Below the convection zone (i.e. within 0.7 solar radii) the Sun rotates as a solid (Gough, 2000; Howe *et al.*, 2000). Yearly variations in the rotation rate at the base of the convection zone are probably linked, in some unknown manner, with the generation mechanism for sunspots (Howe *et al.*, 2000).

According to the dynamo theory of sunspot formation, the Sun's weak polar dipole magnetic field originates at the base of the convection zone. The polar field is distorted and amplified to form a toroidal field by more rapid solar rotation at the equator or by shear at the base of the convection zone (Hoyt and Schatten, 1997; Howe *et al.*, 2000). The toroidal field has buoyant instabilities that rise through the convection zone and emerge as magnetic field loops that are linked to sunspots (see note 1 of Howe *et al.*, 2000). At the end of each Schwabe cycle the toroidal field breaks down and the dipolar magnetic field is re-created, but with a reversed polarity. The resulting 22-year cycle in solar magnetic field polarity is called the Hale cycle (Dicke, 1979).

The Hale cycle is less variable in period than the Schwabe cycle (Berger *et al.*, 1990, Fig. 6.22), but it has been suggested that its period might be 20 rather than 22 years if the cycle is indirectly driven by tidal forces related to Jupiter and Saturn (Hibler and Johnsen, 1979). Variations in the amplitude of the Schwabe cycle lasting 80–90 years are termed the Gleissberg cycle. Both the Schwabe and Gleissberg cycles are correlated with variations in the diameter of the Sun, indicating short- and medium-term variations in energy transfer through the convection zone (Gilliland, 1981). It had been suggested that the variation in length of the Schwabe cycle, linked to the phase of Gleissberg cycle, is indicative of medium-term variations in solar activity (e.g. Friis-Christensen and Lassen, 1991). However, recent satellite data show that this inference is incorrect (Fröhlich and Lean, 1998).

The magnetic fields associated with sunspots affect the solar wind. This consists of charged particles streaming out from the Sun that create an electrical current and an interplanetary magnetic field (Lyon, 2000). The effect of the solar wind on the Earth is linked to an interaction with the geomagnetic field. The number of aurorae varies with the Sunspot Number, but with a lag of two or three years (Hoyt and Schatten, 1997). The impact of the solar wind on Earth's environment varies in relation to the solar magnetic polarity (since the Earth's field does not reverse over periods of decades to centuries). Thus if the solar wind affects the Earth's climate it would be expected to exhibit variations with the 22-year Hale cycle period and perhaps longer (Dicke, 1979; Markson and Muir, 1980; Lockwood *et al.*, 1999).

The solar wind deflects cosmic rays (very high-energy particles originating outside the solar system). Cosmic rays generate ^{14}C and ^{10}Be via collision with atmospheric molecules. Thus the atmospheric concentration of these two short-lived radioactive isotopes varies inversely with Sunspot Number (Stuiver and Quay, 1980; Stuiver and Braziunas, 1989; Beer *et al.*, 1990; Solanki *et al.*, 2000). Measurements of ^{10}Be in ice cores and ^{14}C in tree rings can be used to investigate variations in solar activity over periods of decades to thousands of years in the absence of direct observations (e.g. Finkel and Nishiizumi, 1997; Stuiver *et al.*, 1997). There may be a 400-year cycle in solar activity as indicated by ^{14}C data (Stuiver and Braziunas, 1989). However, there are significant differences in the records of ^{14}C and ^{10}Be over the longer periods perhaps indicating a role for ocean dynamics in influencing atmospheric ^{14}C concentrations (Oeschger and Beer, 1990; Sonett and Finney, 1990; Stuiver *et al.*, 1997). Nevertheless, these data support the inference that the prolonged lack/absence of sunspots in the seventeenth century, known as the Maunder Minimum, was associated with reduced solar activity. The Maunder Minimum corresponded to a time of especially cold weather in

Europe, which was one of the multiple cold phases of the Little Ice Age that helped initiate the belief that solar activity and climate are linked (Eddy, 1976). The idea that millennial-scale climatic cycles (Section 6.8) are driven by long-term solar forcing, based on GISP2 ice-core measurements of ^{10}Be flux (van Geel *et al.*, 1999), has been discussed by the GISP2 workers and discounted as unproven (Finkel and Nishiizumi, 1997; Stuiver *et al.*, 1997).

Connections between solar variability and climate have been pursued for two centuries (reviewed by Burroughs, 1992; Hoyt and Schatten, 1997). In many cases there are spectral analyses demonstrating the Schwabe and/or Hale cycle in meteorological data (e.g. Currie, 1987). Yet in the absence of a convincing physical mechanism to explain these observations in terms of solar forcing, much useful empirical work has been ignored by meteorologists and climatologists. The total measured luminosity variation over an 11-year cycle is just $\pm 0.15\%$, increasing to $\pm 0.25\%$ for monthly periods and $\pm 0.5\%$ over daily periods (Hoyt and Schatten, 1997). These figures relate to the variation for all radiation frequencies. Although the maximum solar emission occurs in the visible spectrum, the maximum proportional change occurs at ultraviolet (UV) frequencies; the UV variation constitutes 9% of the total irradiance, but 32% of the changes. UV variations have virtually no impact on the troposphere because of stratospheric absorption.

However, recent general circulation modelling suggests that UV variations have a significant impact on the stratosphere via direct heating and the UV-induced production of ozone which increases the potential for heating (Haigh, 1996; Robock, 1996; Kerr, 1999; Shindell *et al.*, 1999). The effects of this variable stratospheric heating may affect storm track paths in the troposphere and thus perhaps local precipitation at mid-to-high latitudes (Kerr, 1999; Shindell *et al.*, 1999). This variability is concentrated at the period of the Schwabe cycle (11 years). The modelling currently lacks variable sea-surface temperatures in the oceans, the effect of the Quasi-biennial Oscillation (Labitzke and van Loon, 1990) and transportation of ozone that is interactive with the changing boundary conditions (Shindell *et al.*, 1999). Nevertheless, there does appear to be a successfully modelled link between solar variability and the North Atlantic Oscillation over time scales of decades to centuries (Shindell *et al.*, 2001).

Another effect that has been postulated is a relationship between the interplanetary field and the electric field of the Earth (Markson and Muir, 1980; Tinsley, 1996, 1997). It is argued that since the ionization of the upper atmosphere is partly correlated with the effects of the solar wind, through varying the flux of cosmic rays, perhaps the creation of thunderstorms is linked and has an effect on global weather. More work is needed to understand cloud microphysics before the precise implications of interplanetary field control of the climate can be assessed (Tinsley, 1997). It remains to be seen whether any solar forcing via the Earth's electrical field operates at the period of the Hale cycle as well as the Schwabe cycle period.

Instrumental temperature data, ^{10}Be and δ^{18}O in ice cores, ^{14}C in tree rings, tree-ring widths and densities all point to a connection between solar variability and climate (Stuiver and Braziunas, 1989; Beer *et al.*, 1990; Briffa *et al.*, 1990; Friis-Christensen and Lassen, 1991; Stuiver *et al.*, 1995, 1997; Thomson, 1995; Mann *et al.*, 1998). However, solar forcing is not consistently the most significant factor in climatic variations over time scales of decades to centuries (Thomson, 1995; Mann *et al.*, 1998).

6.7.2   Stratigraphic records of solar cycles

Sedimentological records of the short sunspot cycles (Schwabe and Hale cycles) are not very common (Anderson, 1961; Anderson and Koopmans, 1963; Fischer, 1986). The time series used for their detection are mostly periodic, discrete-signal records (Section 2.2.2) based on the thicknesses of varves. However, Dunbar *et al.* (1994) detected cycles with periods of 83, 23 and 11 years in their long sub-annual coral oxygen-isotope record from the Galapagos Islands which they linked to solar forcing. Yu and Ito (1999) argued, based on decadal-resolution geochemical lake data, for 400-year cycles of drought in North America. Varves in a modern proglacial lake in Alaska also show weak evidence for solar forcing (Sonett and Williams, 1985). Anderson (1992b) argued that early Holocene varves in the USA contain evidence for the Hale (22 year) and perhaps Gleissberg cycles (90 year), but the spectral evidence is not overwhelming. Neff *et al.* (2001) argued for the influence of the Hale cycle on the strength of the Indian summer monsoon based on an early Holocene oxygen-isotope record from a stalagmite in Oman. There is also evidence for 90-year and possibly 11-year (Schwabe) cycles in Late Pleistocene-Holocene varves from the Santa Barbara Basin, California (Schaaf and Thurow, 1997).

The Hale cycle was detected in late Pleistocene varve thickness data from New England (Rittenour *et al.*, 2000). The Eocene Green River Formation of Wyoming, USA provides good evidence for the Schwabe cycle as originally proposed by Bradley (1929; Crowley *et al.*, 1986; Ripepe *et al.*, 1991). Algeo and Woods (1994) have good spectral evidence for regular cycles interpreted as having periods of 22 and 70 years from the Carboniferous. Finally, at one stage the varying thicknesses of Elatina laminites from the Proterozoic were interpreted as a record of sunspot cycles (Williams, 1981; Williams and Sonett, 1985; Williams, 1986). However, subsequent work led to a re-interpretation in terms of tidal control (Williams, 1988, 1989). Note that the relationship between amplitude and cycle period used as evidence for sunspot cyclicity (Williams and Sonett, 1985) is also consistent with tidally mediated cyclicity (Williams, 1989).

In virtually all these cases it is simply the coincidence of the observed regular cycle periods with solar cycle periods which has been used to imply solar forcing. Until a firm connection between solar activity and climatic change has been established, via plausible physical models of the atmosphere, the reason for the variety of apparently solar-cycle periods detected in different sedimentary records cannot be assessed. In fact a non-linear general circulation model, that excluded external forcing, oscillated at a period of 10–12 years (James and James, 1989). This implies that the presence of an 11-year cycle alone in a stratigraphic record is insufficient to claim solar forcing (Burroughs, 1992). Furthermore, decadal variability has been established for the North

Atlantic Oscillation (Section 6.6) at which period no solar forcing is implied (Hurrell, 1995; Sutton and Allen, 1997; Black *et al.*, 1999). This emphasizes the necessity for realistic mechanisms, if a case is to be made for a solar activity origin for cycles in stratigraphic records (e.g. compare Stuiver *et al.*, 1995 with White *et al.*, 1997 and Johnsen *et al.*, 1997, on the subject of the Schwabe cycles in oxygen-isotope variability in the Greenland Summit ice cores, and see Fig. 6.3).

van Geel *et al.* (1999) speculated on a link between solar variability and records of ^{10}Be and climatic variability obtained from ice cores. Recently, Bond *et al.* (2001) compared variations in the production rates of ^{10}Be and ^{14}C with variations in drift ice extent and associated ocean surface circulation changes in the North Atlantic over the last 12,000 years. These variations have similar unique patterns at time scales of hundreds to a few thousand years and may be part of a climate response linked to solar variability extending outside the North Atlantic at least as far as Oman (Bond *et al.*, 2001; Neff *et al.*, 2001). Indeed Bond *et al.* (2001) account for at least the Holocene portion of the quasi-periodic millennial-scale cycles (discussed in the next section) via indirect solar forcing. The exact mechanisms linking solar variability to variations in atmospheric dynamics and the oceanic thermo-haline circulation (discussed in Section 6.8) require further investigations (Bond *et al.*, 2001; cf. Shindell *et al.*, 2001).

6.8 Millennial-scale cycles and Heinrich events

6.8.1 Latest Pleistocene Heinrich events and millennial-scale climatic cycles in the North Atlantic

Recently there has been a great deal of interest in evidence for environmental cycles with periods of between 1000 and 10,000 years (e.g. papers edited by Clark *et al.*, 1999). The time series used are, virtually without exception, based on continuous-signal records (Section 2.2.1). Deep-sea sediments from the North Atlantic contain thin horizons rich in siliciclastic and carbonate-clastic material derived from land by short episodes of ice rafting (Ruddiman, 1977; Heinrich, 1988). The horizons rich in ice-rafted debris (IRD) are indicative of Heinrich events that lasted perhaps less than a thousand years and occurred at intervals of 7–13 ka during the last hundred thousand years (Bond *et al.*, 1992; Broecker *et al.*, 1992). The average period of the Heinrich events is about 7000 years and the youngest example (H-0) is associated with the Younger Dryas cooling episode at around 12 ka BP (Bond *et al.*, 1993; Andrews *et al.*, 1995). Provenance studies indicate that most of the terrigenous particles were transported from the Laurentide ice sheet, emerging through the Hudson Strait west of Greenland during ice-sheet collapse phases (Bond *et al.*, 1993; Gwiazda *et al.*, 1996). Huge flotillas of icebergs (McManus *et al.*, 1998) carried the IRD into the North Atlantic (Bond *et al.*, 1992; Alley and MacAyeal, 1994; Francois and Bacon, 1994; Thomson *et al.*, 1995). This material apparently accumulated synchronously in the Nordic Seas (Fronval *et al.*, 1995; Elliot *et al.*, 2000).

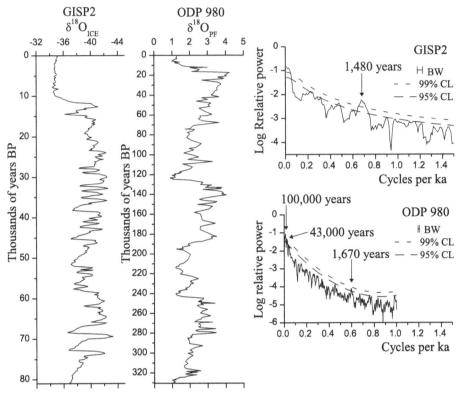

Fig. 6.23 $\delta^{18}O$ from the GISP2 ice core (Fig. 6.1, Table 1.1, sample interval 200 years, see Section 6.2 for source of data) from 0 to 80.8 ka BP and $\delta^{18}O$ from ODP 980 in the North Atlantic (interpolated sample interval 500 years, data courtesy of Jerry McManus) from 0 to 330 ka. Both records show strong evidence for millennial-scale cyclicity (period approximately 1500 years). The millennial-scale cyclicity is superimposed on the dominant 100-ka and 41-ka Milankovitch cyclicity (Section 6.9). BP: before present; $\delta^{18}O_{PF}$: $\delta^{18}O$ from planktonic foraminifera (i.e. marine surface water).

Ice cores from the Greenland ice sheet contain evidence for substantial (5°C) variations in temperature and atmospheric dust concentrations that recurred with periods of a few thousand years between 115 and 10 ka BP (Figs. 6.5 and 6.23, Greenland Ice-core Project Members, 1993; Dansgaard *et al.*, 1993). These variable climatic states, or stadial/interstadial temperature oscillations, over the Greenland ice sheet are referred to as Dansgaard–Oeschger cycles. It is now clear that from 20 to 110 ka BP, the Greenland temperature oscillations coincided with minor IRD events and Heinrich events, as well as changes in sea-surface temperature, with periods close to 1.5 ka (1–3 ka, Fig. 6.23, Bond *et al.*, 1993; McManus *et al.*, 1994; Fronval *et al.*, 1995; Bond *et al.*, 1997; Grootes and Stuiver, 1997). These cycles apparently continue to the present, with the Little Ice Age being the most recent millennial-scale temperature minimum (Bond *et al.*, 1997; Bianchi and McCave, 1999; McManus *et al.*, 1999). Iceberg discharges associated with millennial-scale cycles originate from several sites around the North Atlantic (Bond and Lotti, 1995). Wunsch (2000a) doubted the reality of the ice-core spectral evidence for quasi-periodic millennial cycles in isotopes, believing that the associated spectral peak resulted from aliasing of the annual cycle. However, millennial-scale variations in bottom-water velocity in the North Atlantic cannot result from aliasing since there is no significant annual variation in current speed (Bianchi and McCave, 1999; McCave pers. comm., 2000). Furthermore, recent analyses have shown that aliasing is not a credible explanation for millennial-scale spectral peaks associated with the ice-core records (Alley *et al.*, 2001; Meeker *et al.*, 2001).

6.8.2   Climatic mechanisms involved in North Atlantic millennial-scale cycles and Heinrich events

The Salt-Oscillator model of Broecker and Denton (1990) and Broecker *et al.* (1990) involves variations in the vigour of oceanic thermohaline circulation (or 'ocean conveyor') as driven by bottom-water formation in the North Atlantic. This model can account for short-period (<10 ka) climatic changes at least in the North Atlantic region (e.g. Ganopolski and Rahmstorf, 2001, 2002; Clark *et al.*, 2002). The rapid input of fresh water to the North Atlantic during Heinrich-type iceberg discharges, or the release of melt water from ice-damned lakes and melting ice sheets lowered surface water density so preventing formation of the North Atlantic Deep Water (NADW, Schmitz and McCarthey, 1993; Seidov and Maslin, 1999). During the absence or reduction in bottom-water formation, whether during millennial-scale cycles or Heinrich events, there is a considerable reduction of heat transfer from the Gulf Stream/North Atlantic Current to northern Europe and hence regional cooling. Supporting the Salt-Oscillator model, stratigraphic records show that millennial-scale cycles are associated with changes in the production of NADW (Keigwin and Lehman, 1994; Oppo and Lehman, 1995; Bianchi and McCave, 1999). It appears likely that the mechanism, locus and intensity of bottom-water formation all varied throughout the Late Pleistocene (e.g. Dokken and Jansen, 1999; McManus *et al.*, 1999).

The Heinrich events do not involve exceptional circulation patterns (Oppo and Lehman, 1995). Instead they apparently represent unusually intense and prolonged cases of every third to fifth Dansgaard–Oeschger or millennial-scale cycle (Bond *et al.*, 1993). The variation in the intensity of the millennial-scale cycles, i.e. from Heinrich event to Heinrich event, is sometimes described as the Bond cycle. Heinrich events have been modelled using a 'Binge–Purge' model of the Laurentide Ice Sheet (MacAyeal, 1993a, 1993b; Alley and MacAyeal, 1994). This accounts for exceptionally large iceberg discharges from the Hudson Strait with a period of around 7000 years without requiring an external periodic driving mechanism. However, it does not explain the way in which Heinrich events fit into Bond cycles as the culmination of millennial-scale cycle oscillations (Bond *et al.*, 1993). Hunt and Malin (1998) proposed that earthquakes caused by ice-loading under the Laurentide Ice Sheet were responsible for catastrophic ice-sheet collapse and iceberg discharge to the North Atlantic. Their model has the merit of explaining the variable period between Heinrich events, but not the relationship with minor millennial-scale cycles.

However, Alley *et al.* (2001) argue that low-amplitude periodic (1.5 ka) forcing of a noisy system exhibiting stochastic resonance (Benzi *et al.*, 1982; Wiesenfield and Moss, 1995; Bezrukov and Vodyanoy, 1997) would account for the observed variability in the amplitude and period of the millennial-scale cycles in the North Atlantic. Systems exhibiting stochastic resonance show a non-linear response above a certain threshold. This means that whereas a weak periodic forcing of the system might be undetectable on its own, the addition of random noise can mean that the periodic signal becomes detectable (amplified). Recently, Ganopolski and Rahmstorf (2002) were able to generate stochastic resonance in a general circulation model that has the same characteristics as the Dansgaard–Oeschger cycles in the GRIP and GISP2 oxygen-isotope records, by using realistic variability in freshwater fluxes to the North Atlantic. Additionally, Bond *et al.* (2001) showed that in the Holocene solar variability and drift ice extent in the North Atlantic were probably linked – suggesting a possible solar origin for millennial-scale cyclicity.

6.8.3   Evidence for millennial-scale cycles outside the North Atlantic region

Relevant to the ultimate cause of the millennial-scale cycles are the indications of their occurrence outside the North Atlantic region. In fact there is growing evidence for the worldwide expression of the millennial-scale cycles. In the equatorial Atlantic, varying surface- and bottom-water conditions are certainly linked to NADW production and probably trade-wind intensities (Curry and Oppo, 1997; Arz *et al.*, 1998). This is reminiscent of the link between the Atlantic off Venezuela and the North Atlantic Oscillation (Section 6.6, Chang *et al.*, 1997; Black *et al.*, 1999). On land, the climatic history of North African and South American lakes was also linked to North Atlantic variations in the ocean conveyor (Street-Perrott and Perrott, 1990; Baker *et al.*, 2001). Ninnemann *et al.* (1999) demonstrated millennial-scale changes in ocean circulation in the Southern Ocean.

Following the thermohaline circulation eastwards, in the northern Indian Ocean (the Arabian Sea) variations in the burial of organic carbon and formation of alternating laminated/bioturbated sediments correlates with the stadial/interstadial conditions in the North Atlantic (Schulz *et al.*, 1998). Changing bottom-water oxygenation may relate to both varying oceanic circulation and to productivity that is indirectly linked to upwelling controlled by the Indian monsoon (Schulz *et al.*, 1998).

Isotopic evidence for millennial-scale ocean circulation changes also exists in the Northeast Pacific (Lund and Mix, 1998) and North Pacific iceberg discharges might correlate with the Dansgaard–Oeschger cycles (Kotilainen and Shackleton, 1995). In the Santa Barbara Basin off California, a similar situation to the Arabian Sea exists with bioturbated sediments deposited during stadials and less-bioturbated to laminated sediments preserved during interstadials (Behl and Kennett, 1996). Changing bottom-water conditions and surface productivity produced pronounced variations in the populations of benthic foraminifera (Cannariato *et al.*, 1999). Variations in bottom-water oxygenation were linked to the changing direction of ocean circulation (Kennett and Ingram, 1995) and the depth of the thermocline (Hendy and Kennett, 1999).

The abrupt (decadal) switches in sea-surface and bottom-water conditions in the Santa Barbara Basin have been used to argue that atmospheric reorganizations, such as changes in the position of the jet stream, rather than just oceanic processes account for the observations (Hendy and Kennett, 1999). Another atmospheric connection to millennial-scale cycles comes from consideration of greenhouse gases such as water vapour, carbon dioxide and methane which might all be linked to ocean circulation changes. Methane concentrations in ice bubbles in the Byrd ice core of Antarctica and the GRIP and GISP2 cores from Greenland vary at stadial/interstadial time scales by about 100 ppb (Bender *et al.*, 1994; Brook *et al.*, 1994; Stauffer *et al.*, 1998). This is enough to cause a 0.3°C variation in temperature over Antarctica. Carbon dioxide concentration in ice from Antarctica, where the record is less affected than in Greenland by gas evolved from detrital carbonate dust, varies in phase with the Heinrich events by about 30 ppm, corresponding to a 0.6°C variation in temperature (Stauffer *et al.* 1998). It is not clear whether the response time of carbon dioxide in the oceans is too slow for millennial-scale variations to be detected (Stauffer *et al.*, 1998). Combined, the methane and carbon dioxide changes can explain about 20% to 30% of the stadial/interstadial temperature changes. However, variations in water vapour could have contributed to varying atmospheric temperatures, but this is hard to assess from the stratigraphic record (Broecker, 1997).

6.8.4   Earlier records and the origin of millennial-scale cycles

The record of millennial-scale cycles has been extended back from the latest Pleistocene by McManus *et al.* (1999, Fig. 6.23) using sediments from ODP Site 980 on the Feni Drift in the North Atlantic (Fig. 6.1). Their results indicated that Heinrich-event-type pulses of IRD had occurred in glacial phases and stadials throughout the interval from 0.1 to 0.5 million years ago. No significant IRD events occurred when global ice volumes were small enough that benthic $\delta^{18}O$ was less than 3.5‰ PDB. SST variability at stadial/interstadial time scales is much greater during glacial times. However, bottom-water productivity variations occurred even during the warmest interglacial and interstadial times. This suggests that production of NADW is not entirely linked to ice-sheet volumes.

Raymo *et al.* (1998) used measurements of benthic foraminiferal $\delta^{18}O$ and IRD to infer millennial-scale cycles at 1.2–1.4 Ma BP in the North Atlantic. However, there is good power-spectral evidence for millennial-scale cycles throughout the Palaeozoic. Thus evaporite records from the Ordovician–Silurian (Williams, 1991) and Permian (Anderson, 1982) and cycles in re-deposited carbonate from shallow continental shelf settings in the Cambrian, Devonian and Carboniferous (Elrick *et al.*, 1991; Elrick and Hinnov, 1996) point to the temporal prevalence of such short-term climatic variability.

The regularity, global presence and temporal persistence of millennial-scale cycles apparently indicates some significant external forcing. It has been suggested repeatedly that climatic variability at periods of 10 ka and less arise from harmonics and/or combination tones (Section 5.2.4) of Milankovitch cycles (i.e. a non-linear response, Le Treut and Ghil, 1983; Pestiaux *et al.*, 1987; Short *et al.*, 1991; Hagelberg *et al.*, 1994). This would explain the persistence and regularity of millennial-scale cycles, but not their comparatively large amplitude given their period. Thus it would be expected that the cycles with periods of about 1.5 ka would occur with longer period components having larger amplitudes. In the case of harmonics of the precession cycle, the 7-ka Heinrich events might represent the third harmonic, but the 1.5-ka period would represent the fourteenth and thus be expected to be of very small amplitude. Clearly some model of non-linear amplification would be required to account for the relative strength of the third and fourteenth harmonics and the weakness of the other harmonic components of the precession cycle.

A different orbital control would arise from the 1.6-ka tidal cycle when conjunction/ opposition, perigee and perihelion coincide (Section 6.3). This tidal cycle has recently been invoked as an explanation for millennial-scale cycles, though the climatic mechanisms involved require further study (Keeling and Whorf, 2000; Wunsch 2000b; Munk *et al.*, 2002).

A solar origin for millennial-scale cycles has been postulated given 2.3-ka variations in the production of ^{14}C (e.g. Mayewski *et al.*, 1997; van Geel *et al.*, 1999). However, the measured concentration of ^{14}C in tree rings might actually have been controlled by oceanic circulation, so the suggestion is currently unproven (Finkel and Nishiizumi, 1997; Stuiver *et al.*, 1997). Nevertheless, the similarity of variations in ^{10}Be and ^{14}C production rates to the extent of North Atlantic drift-ice in the North Atlantic during the Holocene strengthens this idea (Bond *et al.*, 2001). Since prediction of long-term climatic change requires an understanding of millennial-scale variability this issue will remain a cause of great interest for some time.

6.9   Milankovitch cycles

6.9.1   The nature and climatic expression of the orbital cycles

The three orbital cycles usually credited with causing climate changes over tens of thousands of years are precession, obliquity, and eccentricity (Figs. 6.24 to 6.27). These orbital cycles arise from the changing gravitational environment in which the Earth orbits the Sun as dictated by the Moon and planets. The changing orbital configuration influences the amount of radiation received from the Sun (the insolation), by differing amounts at different latitudes. Useful introductions to the so-called Milankovitch Theory of climate change are provided by Imbrie and Imbrie (1979) and Berger (1988) and Imbrie *et al.* (1993b). The orbital cycles influence climate by, for example, controlling whether temperatures exceed certain thresholds. For example,

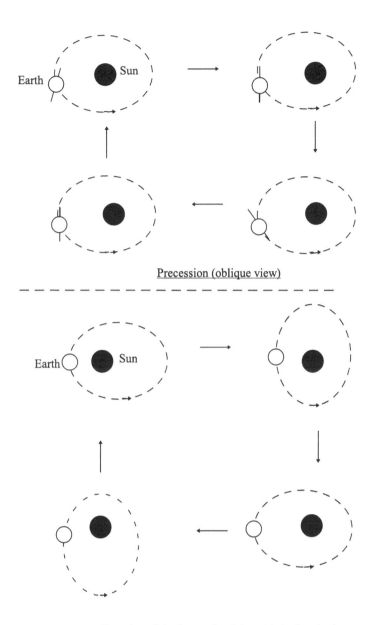

Precession (oblique view)

Rotation of the long axis of the orbit (polar view)

Fig. 6.24 Climatic precession is controlled by two factors: (a) the rotation of the direction of tilt of the Earth's axis and (b) the rotation of the long axis of the Earth's orbit.

Top – Looking down on the northern hemisphere, the Earth moves in its orbit anticlockwise around the Sun. However, the direction of tilt of the Earth's orbit rotates clockwise with a period of 26,000 years relative to the stars. In the figure the thicker part of the axis indicates the axis tilted towards the reader. The figure shows *top left* – tilt to the right, *top right* – tilt away from the reader, *bottom right* – tilt to the left, *bottom left* – tilt towards the reader.

Bottom – The long axis of the Earth's eccentric orbit rotates slowly anticlockwise relative to the stars. Combined with the changing direction of tilt, this leads to the climatic precession, in other words the effect of these two factors on the insolation distribution on the Earth. Climatic precession has a modern period of around 21,000 years. Not to scale.

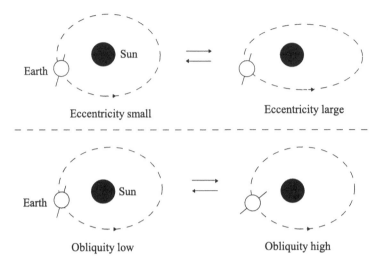

Fig. 6.25 Top – the eccentricity of the orbit varies slightly from almost circular (small eccentricity) to more elliptical (relatively large eccentricity). The eccentricity varies predominantly at periods of 100,000 and 413,000 years. Bottom – the angle of the tilt of the Earth's axis, relative to a direction at right angles to the ecliptic (i.e. the obliquity), varies from less tilted to more tilted and back with a modern period of about 41,000 years. Not to scale.

Milankovitch (1941, cf. Imbrie and Imbrie, 1979) argued that in the Pleistocene–Recent summer temperatures at high latitudes were critical in determining whether all the winter snow and ice melted and thus whether ice sheets grew or contracted. Such orbital cycles influence other planets; it is likely that the polar layered deposits on Mars record 'orbital-forcing' of climate (Murray *et al.*, 1973; Toon *et al.*, 1980).

The idea that changes in the Earth's orbital geometry have influenced climate and might be of importance to geology is not new. Lyell (1830, p. 110) stated:

> Before the amount of difference between the temperature of the two hemispheres was ascertained, it was referred by astronomers to the acceleration of the earth's motion in its perihelium; in consequence of which the spring and summer of the southern hemisphere are shorter, by nearly eight days, than those seasons north of the equator. A sensible effect is probably produced by this source of disturbance, but it is quite inadequate to explain the whole phenomenon. It is, however, of importance to the geologist to bear in mind, that in consequence of the procession [sic] of the equinoxes the two hemispheres receive alternately, each for a period of upwards of 10,000 years, a greater share of solar light and heat.

Thus Lyell drew attention to the potential geological significance of the precession cycle (Fig. 6.24). However, Herschel (1832) pointed out that the total annual insolation at any point on the Earth's surface does not vary at the precession period. He presented a discussion to the Geological Society of London concerning the eccentricity and obliquity cycles as well, but discounted the latter as being too small in amplitude to be of importance for climate variation. Imbrie and Imbrie (1979) provided a highly readable

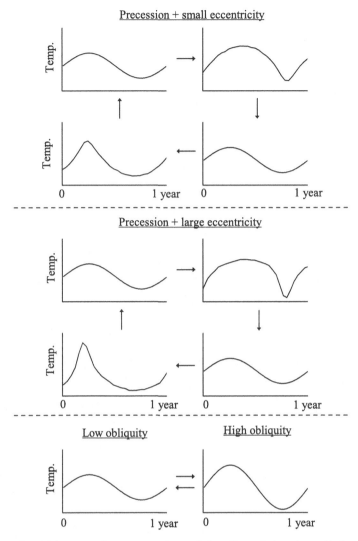

Fig. 6.26 Schematic representations of the effect of changing orbital parameters on the temperature at mid latitudes – ignoring regional geographical and climatic factors (cf. Short *et al.*, 1991). The figures have the seasons aligned for ease of comparison. Top – climatic precession causes a small change in the relative length and duration of the seasons. Top and middle – the size of the precession effect depends directly on the size of the eccentricity. Bottom – the obliquity cycle controls the degree of seasonality of climates. Not to scale.

account of the history between 1842 and the mid 1970s of the orbital-climatic theory as an explanation for the Pleistocene ice ages. They drew attention to the fact that it was Croll (e.g. 1864) in the 1860s who formulated the first credible orbital-climatic hypothesis.

Nevertheless, Milankovitch's (1941) considerable efforts to quantify insolation variations and his pursuit of orbital-climatic connections led to the theory that bears

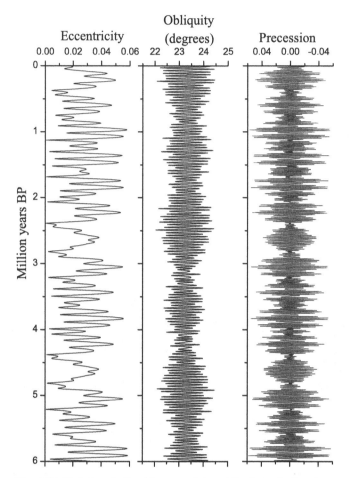

Fig. 6.27 Variations in Earth's eccentricity, obliquity and climatic precession over the last 6 million years according to the (1, 1) orbital solution of Laskar *et al.* (1993a).

his name. However, his theory was not widely adopted until the classic paper by Hays *et al.* (1976). Their spectral analysis of benthic foraminiferal oxygen-isotopes as an indication of global ice volume, plus sea-surface temperature estimates from southern hemisphere radiolarian assemblages, established beyond doubt that the late Pleistocene climate changes had the same periods as the orbital cycles.

6.9.1a   Precession

Orbital precession involves two factors that relate to: (a) the direction of tilt of the Earth relative to the stars, and (b) the orientation of the eccentric orbit relative to the stars (Fig. 6.24). The direction of tilt of the Earth rotates in space, or precesses, so that the northern and southern hemispheres experience summer alternately during the closest approach to the Sun (perihelion). This general precession has a period of 26 ka with a clockwise direction as viewed looking down on the North Pole. However, simultaneously, the long axis of the eccentric orbit of the Earth rotates anticlockwise relative to the stars. Thus in each hemisphere, 11 ka after summer occurs during perihelion, summer occurs during aphelion (the furthest point from the Sun). The combined effect of these two orbital factors, with a period of approximately 21 ka (Figs. 4.21 and 5.18), is sometimes called 'climatic precession' – referring to the planet-based rather than astronomical point of view.

The Earth speeds up in its orbit when it is closer to the Sun. For a particular hemisphere, the hotter days of summer near perihelion are balanced by a larger number of cooler days near aphelion. Consequently, the total insolation during a year is unaffected by whether a particular hemisphere has summer during perihelion or during aphelion. Overall for each hemisphere (but with the hemispheres 180° out of phase), the climate at medium-to-high latitudes progresses as: (a) medium summer/medium winter, (b) short hot summer/long cool winter, (c) medium summer/medium winter, and (d) long warm summer/short cold winter (Fig. 6.26). The changing length of the seasons in relation to precession partly accounts for the non-sinusoidal shape of annual records of temperature and the resulting harmonic in the spectrum of central England temperature (Section 5.2.4, Fig. 6.7). As just explained, there is no annual variation in total insolation due to precession. However, variations in seasonal insolation due to precession are larger at low latitudes than variations due to obliquity, and comparable to obliquity at the poles (Imbrie *et al.*, 1993b).

Since precession north and south of the equator is 180° out of phase, the simultaneous changes in climates in the northern and southern hemispheres at periods of around 20,000 years, demonstrated by Hays *et al.* (1976), was initially seen as a problem for the Milankovitch theory (e.g. Imbrie and Imbrie, 1979). However, it now appears likely that factors such as sea level and atmospheric carbon dioxide concentrations (and hence temperature) can explain forcing of the world's climate at the pace of the currently dominant northern-hemisphere ice sheets (e.g. Shackleton & Pisias, 1985; Shackleton, 2000).

6.9.1b   Eccentricity

The precession effect arises because of the elliptical nature of the Earth's orbit. The different intensity and length of the seasons during the precession cycle are caused by the changing Earth–Sun distance and the changing speed of the Earth along its orbit. This means that changes in eccentricity have a direct effect in modulating the amplitude of the precession cycle (Figs. 6.25 to 6.27). Eccentricity varies, particularly at periods of about 100 ka (more accurately 95 and 125 ka) and approximately 400 ka. These multiple frequency components lead to the characteristic heterodyne amplitude modulation (Section 1.3) of precession (Fig. 5.18). There are additional less well known and smaller amplitude longer-term variations in eccentricity that modulate the precession cycle, most notably with a modern period of about 2.4 million years, that may have been identified in stratigraphic records (e.g. Hilgen *et al.*, 1995; Olsen and Kent, 1999). The eccentricity of the Earth's orbit is very small and the changes are small so there is an almost negligible influence of eccentricity variations on total annual insolation (less than 0.2%, p. 638 of Berger, 1988; Imbrie *et al.*, 1993b). Nevertheless, the apparently large amplitude of changes in global ice volumes in the late Pleistocene at the frequency of the 100-ka eccentricity cycle have led to extensive investigations and speculations (Section 6.9.4b).

6.9.1c   Obliquity

The tilt or obliquity of the Earth's orbit stays within relatively narrow limits (about 22–24.5°) compared to planets such as Mars, due to the stabilizing effect of the Moon (e.g. Laskar *et al.*, 1993b). The dominant period is 41 ka (Figs. 4.20, 4.21, 6.25, 6.26 and 6.27), but it has a 1.25-million-year amplitude modulation (e.g. Shackleton *et al.*, 1999a). The larger the tilt, the larger the amplitude of the annual cycle since the annual or seasonal cycle is a direct result of the tilt. Thus, obliquity causes imposed amplitude modulation (Section 1.3) of the annual cycle. The difference in period between the annual cycle and the obliquity period is very large. Thus in spectra of records of the annual cycle, the combination tone peaks related to this imposed amplitude modulation (at frequencies of 1/1 year + 1/41,000 year and at 1/1 year − 1/41,000 year) cannot be distinguished from the annual spectral peak (Fig. 6.7). Variations in tilt naturally lead to variations in the degree of seasonality with the same timing (phase) in the northern and southern hemispheres. Annual changes in insolation related to obliquity are greatest at high latitudes. However, seasonal changes in insolation due to obliquity and precession are comparable at high latitudes (Imbrie *et al.*, 1993b). It had been speculated that obliquity influences the geomagnetic intensity (reviewed by Kent, 1999). However, recent long high-resolution records of intensity are not coherent with any of the orbital periods, and geomagnetic reversals and intensity variations can be explained by interactions between Earth's core and the mantle thermal structure (Glatzmaier *et al.*, 1999; Guyodo and Valet, 1999).

6.9.2   Earth's orbital history

The motion of the planets is 'chaotic' (Section 4.7) meaning that inevitably the trajectories of the calculated orbital histories diverge from the correct history at some point in the past (Laskar, 1989). This follows because the initial starting conditions, or the current astronomical conditions, cannot be specified with infinite accuracy and infinite precision. It currently appears that the orbital solution (e.g. Laskar *et al.*, 1993a) cannot be reliably extended much further than the mid-to-late Eocene or around 35–50 Ma BP (Laskar, 1999). However, the frequencies of the individual orbital cycles can be relied upon for earlier times, given an important caveat. In other words, the phase relationship between different orbital cycles cannot be calculated for say the Mesozoic, but the periods of the orbital cycles can be deduced.

The caveat is that the average periods of the precession and obliquity cycles are not fixed. The gradually increasing Earth–Moon distance, partly related to tidal dissipation, has meant that both these orbital cycles have lengthened at the rate of a few percent per 100 million years (Berger *et al.*, 1989). This very slow rate of change means that, for the analysis of stratigraphic records spanning as much as tens of millions of years, the average periods of precession and obliquity can be considered to be virtually constant. The calculated history of the precession and obliquity periods was derived using length of day information from banding in fossil invertebrates (Berger *et al.*, 1989). However, more recent tidal cycle studies will probably prompt a revision of the calculations (Section 6.3.3).

The dynamical ellipticity of the Earth, affected by changing ice-mass volumes and distributions and mantle convection, also affects the history of orbital changes and hence insolation history (e.g. Dehant *et al.*, 1990; Rubincam, 1995; Forte and Mitrovica, 1997; Mitrovica *et al.*, 1997). However, despite the uncertainties in the history of both tidal dissipation and dynamical ellipticity, cyclostratigraphic studies of stratigraphic records of Milankovitch cycles are currently helping to clarify the values of parameters used in the calculations of orbital history (Lourens *et al.*, 1996; Olsen and Kent, 1999; Lourens *et al.*, 2001; Pälike and Shackleton, 2000).

6.9.3 Orbital tuning

6.9.3a Tuning of Pliocene–Recent cyclostratigraphic records

The Hays *et al.* (1976) paper employed orbital tuning to refine the time scales for their time series. Band-pass filtering (Section 4.3) of the records using the initial time scale allowed comparison of precession-scale and obliquity-scale oscillations filtered from the data, with the orbital history of precession and obliquity. Tuning involved alignment of the filtered records with the calculated orbital history using constant phase lags (since at these frequencies climate takes a few thousand years to respond to the changes in insolation). A similar strategy was adopted by Imbrie *et al.* (1984) who produced a stacked record of benthic foraminiferal oxygen isotopes from five deep-sea cores (each standardized by subtracting the means and dividing by the standard deviations). Their time series of standardized oxygen isotopes, known as the SPECMAP stack, was tuned to the orbits back to 780,000 years ago. From the 1980s the SPECMAP time series was widely used as a template for dating late-Pleistocene deep-water oxygen-isotope records.

Before tuning is used it is important to show that there is spectral evidence for regular cyclicity (e.g. Ruddiman *et al.*, 1989; Rial, 1999, Fig 6.28). Pisias (1983) argued that tuning random numbers does not lead to spuriously increased coherency (Section 4.5.1) between the tuned record and the tuning target in the tuning frequency band. Thus increasing coherency is often considered to be a sign of successful tuning. Tuning has also been judged successful when there was an observed increase in coherency in the orbital-frequency bands that were not tuned (Imbrie *et al.*, 1984). Martinson *et al.* (1982) provided an automated method for matching a cyclostratigraphic record to a target orbital history by distorting the time–depth relationship, or 'mapping', so as to maximize coherency.

However, Shackleton *et al.* (1995c) showed that caution is required when coherency alone is used to judge the success of tuning. Increased coherency at a particular orbital

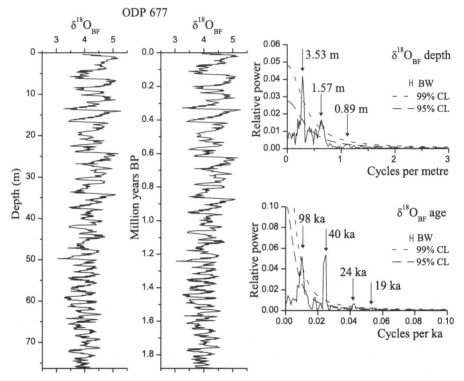

Fig. 6.28 $\delta^{18}O$ from benthic foraminifera (= BF) from ODP Site 677 in the eastern equatorial Pacific (Fig. 6.1, Table 1.1) plotted as a function of depth (0–76.3 m) and tuned age (0–1.881 Ma BP, Shackleton *et al.*, 1990, see Section 6.2 for source of data). The ODP 677 oxygen isotopes show three regular cycles as a function of depth. Note that many authors prefer to plot $\delta^{18}O$ with a reversed scale (i.e. with lower $\delta^{18}O$, or times of minimum ice volumes, on the right rather than, as here, on the left).

frequency can arise when there is alignment of the record to the wrong set of orbital cycles. Furthermore, if a cyclostratigraphic record is tuned at one frequency to the wrong part of the orbital history, this can lead to increased coherency in other orbital frequency bands. Thus complex demodulation (Section 4.4) should be used to check that the amplitude modulation of the tuned data matches (is in phase with) the amplitude modulation of the target curve.

Orbital tuning has usually involved the use of a constant phase relationship or time lag between the environmental variable represented by the cyclostratigraphic record and the orbital history. However, Pisias *et al.* (1990) demonstrated that, at least over the last million years, non-linearities in the climate system involved variable response times. Hence the time lag between the orbital changes and climatically driven variations was probably not perfectly constant. For most purposes the variation in lag between orbital changes and the climate is probably not significant, although the phase relationships between different parts of the climate system appear to have changed over the last three million years (Clemens *et al.*, 1996; Rutherford and D'Hondt, 2000).

Shackleton *et al.* (1990) tuned oxygen isotopes at ODP Site 677 in the eastern equatorial Pacific. Previous workers started by examining their records using a time scale established using the published ages of the geomagnetic reversals (e.g. Ruddiman *et al.*, 1989). Instead, Shackleton *et al.* (1990) started by isolating the oscillations relevant for tuning by band-pass filtering in the depth domain, as there was clear evidence for very regular cyclicity (Fig. 6.28). Their tuning involved the use of precession-related cycles in planktonic foraminiferal oxygen isotopes that were partly controlled by local sea-surface temperature. The benthic oxygen isotope variations from Site 677, believed to relate mainly to global ice volume, were dominated by obliquity-related cycles before about 1.3 Ma BP. As a global signal, the Site 677 benthic foraminiferal isotopes correlated with those at Deep Sea Drilling Project (DSDP) Site 607. However, the new tuning implied that the tuning at Site 607 by Ruddiman *et al.* (1989) had incorrectly identified a few cycles. Furthermore, the tuning at Site 677 implied that the generally accepted radiometric dates for the geomagnetic reversals were too young by around 5% to 7%. Their result confirmed the observation by Hilgen and Langereis (1989) that there was a problem with the time scale. The new tuning produced more linear age/depth plots than the age models used by Ruddiman *et al.* (1989) or the SPECMAP group.

This was a critical step in the history of orbital tuning. Subsequently, tuning from the Recent back into the Pliocene proceeded without total reliance on the radiometrically dated reversal ages (e.g. Hilgen *et al.*, 1995; Shackleton *et al.*, 1995a). The radiometric dating had used K-Ar dates, but Ar-Ar dating of key sections recording geomagnetic reversals confirmed that the astronomically tuned dates were more reliable (see review by Kent, 1999). The tuned age model for the Pliocene to Recent reversals was subsequently adopted for sea-floor spreading models and in a revised geomagnetic polarity time scale (Wilson, 1993; Cande and Kent, 1995; Kent, 1999).

6.9.3b   Tuning of older cyclostratigraphic records

Orbital tuning of pre-Pliocene strata has proceeded using two main methods. For tuning in the mid-to-late Cenozoic, tuning is conducted using the orbital solution as a tuning target. However, it is not yet clear whether tuning in the pre-Pleistocene should treat deep-water oxygen-isotope signals as globally synchronous or not, since tuning alternative variables to the orbital history (e.g. carbonate contents, magnetic susceptibility, colour reflectance) sometimes produces different phase relationships (Clemens, 1999). For older strata, tuning is often conducted using a sine wave as a target for a particular frequency. This can help minimize accumulation rate variations and produce a 'floating' interval time scale that is not 'anchored' to the Recent.

Tuning older stratigraphic cycles to a sine wave in the absence of an orbital solution, or just counting the regular cycles (after their demonstration using spectral analysis), has proved to be a very popular strategy for refining the geological time scale (Gilbert, 1895; Barrell, 1917; House, 1985; collected papers edited by House and Gale, 1995; Shackleton *et al.*, 1999b). Weedon *et al.* (1997) called the estimation of the duration of particular stratigraphic units (e.g. biozones) 'interval dating' to emphasize the distinction from 'relative dating' and 'absolute dating'. However, the obvious problem with studies of ancient sedimentary cycles, in the absence of accurate radiometric dates and an orbital-tuning target, is the problem of potentially undetected stratigraphic gaps (Section 5.3.3).

An example is provided by the Belemnite Marls carbonate data used throughout this book (Section 1.3, Fig. 1.8, Table 1.1). Jones *et al.* (1994) generated a marine $^{87}Sr/^{86}Sr$ curve from belemnite guards some of which were collected from the same Belemnite Marls section analysed by Weedon and Jenkyns (1999). They plotted their strontium-isotope ratios assuming that ammonite subzones were equal in duration. Their plot showed a nearly linear decrease during the first three stages of the Jurassic. Strontium isotopes would not necessarily be expected to change linearly, but in some parts of the marine record the changes in short segments are approximately linear (e.g. Miller *et al.*, 1991; cf. Martin *et al.*, 1999). Using the assumption of Jones *et al.* (1994), Fig. 6.29 shows a linear decrease in strontium isotope values, but also that the constant duration for ammonite subzones implies extremely large variations in the periods of the sedimentary (%CaCO$_3$) cycles.

On the other hand, Weedon and Jenkyns (1999) believed that the main cyclicity could be accounted for by precession cycles, despite the presence of substantially

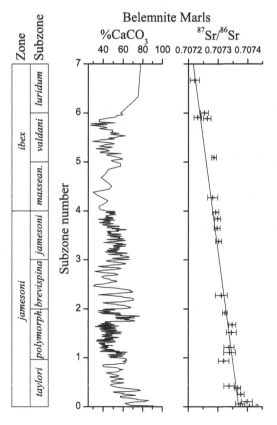

Fig. 6.29 %CaCO$_3$ and $^{87}Sr/^{86}Sr$ from the Belemnite Marls, Dorset, England plotted assuming equal durations for ammonite subzones (data introduced in Section 1.3, Fig. 1.8, Table 1.1). The horizontal bars for the plot of $^{87}Sr/^{86}Sr$ values indicate the 95% confidence intervals. Zone and subzone refers to ammonite biostratigraphy. *massean*: *masseanum*; *polymorph*: *polymorphus*.

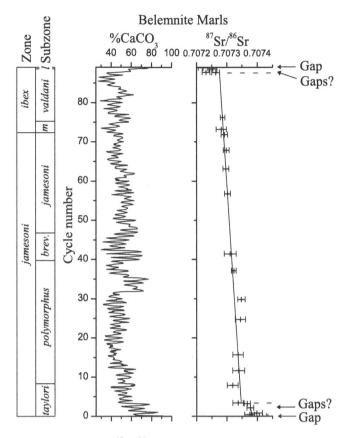

Fig. 6.30 %CaCO$_3$ and ^{87}Sr/^{86}Sr from the Belemnite Marls plotted assuming equal durations for the couplet cycles. There are documented stratigraphic gaps in the top-most and bottom-most beds of the Belemnite Marls (see references in Weedon and Jenkyns, 1999). If marine ^{87}Sr/^{86}Sr decreased linearly during the early Jurassic, the deviation in strontium-isotope ratios from the fitted line near the top and base of the formation implies the presence of undetected stratigraphic gaps. See Fig. 6.29 for ammonite subzone abbreviations.

thinner, though regular, cycles in the top third of the Belemnite Marls (Figs. 4.1 and 4.2). In other words, the thinner cycles implied a decrease in accumulation rate at the top of the section. Plotting these data, assuming equal cycle duration, produced a plot with a linear decrease in ^{87}Sr/^{86}Sr except at the top and base of the unit (Fig. 6.30). However, the thin limestone beds at the very base and the very top of the Belemnite Marls have documented biostratigraphic gaps (see references in Weedon and Jenkyns, 1999). Thus the deviations of the strontium-isotope curve from the linear segment are partly explained by suggesting undetected stratigraphic gaps near both the base and top of the formation near to the stratigraphic gaps that had already been documented (Fig. 6.30). For the majority of the Belemnite Marls a floating time scale was established by fixing the cycle boundaries at 20-ka intervals, allowing the durations of individual ammonite subzones to be determined (Weedon and Jenkyns, 1999). Assuming that the strontium isotope decrease was essentially linear over periods of millions of years, and that this rate had been accurately measured in the bulk of the Belemnite Marls, the durations of the first three stages in the Jurassic were also estimated.

Tuning of late Cenozoic records to the calculated history of insolation variation has allowed considerable improvements in the geological time scale, particularly in terms of the absolute ages of biostratigraphic and magnetostratigraphic events (e.g. papers in Shackleton *et al.*, 1999b). The short time intervals between the selected tie points, often just thousands or tens of thousands of years apart, allows calculation of accumulation rates and, when combined with sediment density, sediment fluxes. In some cases the tuning to one orbital frequency reveals regular components related to other orbital cycles that were not suspected based on analysis in the depth domain. Tuned records are also much easier to correlate to other tuned records, permitting regional or even global studies of a variety of physical and biological processes.

Tuning in the absence of an orbital history target for strata pre-dating the Late Cenozoic also yields accumulation rates and sometimes aids the detection of previously unsuspected regular components. Without the firm absolute dates upon which to fix the resulting 'floating' chronologies, traditional methods are required for correlation (e.g. biostratigraphic and magnetostratigraphic events). Tuning without an orbital history target still yields invaluable information about the time between particular events (interval dating). However, it is hoped that the calculation of the history of the particularly stable approximately 400-ka eccentricity cycle might eventually form a template for absolute dating of some of the otherwise 'floating' tuned records (Preface in Shackleton *et al.*, 1999b).

6.9.4   Stratigraphic records of Milankovitch cycles

6.9.4a   Results from stratigraphic studies

In most cases Milankovitch cyclicity has been identified using continuous-signal records (Sections 2.2.1 and 2.3.3a). However, when pelagic accumulation rates are nearly constant it has proved possible to generate quasi-periodic discrete-signal records permitting FM analysis (Sections 2.2.2 and 2.3.3b). Starting with Gilbert (1895) there is a vast literature concerning stratigraphic records of Milankovitch cyclicity (Fischer, 1980). Reviews are provided by Berger (1989), Fischer *et al.* (1990), Schwarzacher (1993), Weedon (1993) and Hinnov (2000). Collected modern case studies have been edited by, for example, Berger *et al.* (1984), Fischer and Bottjer, (1991), de Boer and Smith (1994), House and Gale (1995) and Shackleton *et al.*, (1999b). A very large number of facies and every period of the Phanerozoic have yielded good evidence for Milankovitch cyclicity backed-up by time-series analysis (Weedon, 1993). Work in the 1980s and early 1990s was devoted to trying to assess how ancient climates were able to generate stratigraphic cyclicity and the problems of signal distortion (Chapter 5). However, much current work on Mesozoic sections is being devoted to improvements of the geological time scale by tuning in the absence of an orbital solution.

One of the most important observations from these stratigraphic records is that at different parts of Earth history, different orbital cycles have dominated the climatic response. Thus in the late Pleistocene, the 100-ka cycles are dominant whereas in the early Pleistocene and Pliocene deep-water oxygen-isotope records are dominated by obliquity cycles (Fig. 6.31, Hays *et al.*, 1976; Imbrie *et al.*, 1984; Ruddiman *et al.*, 1986; Hilgen and Langereis, 1988, 1989; Raymo *et al.*, 1989; Ruddiman *et al.*, 1989; Hilgen, 1991). Late-Miocene records contain strong precession components (Hilgen, 1991; Krijgsman *et al.*, 1994a, 1995; Hilgen *et al.*, 1995; Shackleton *et al.*, 1995a; Tiedemann and Franz, 1997; Shackleton and Crowhurst, 1997; Hilgen *et al.*, 1999).

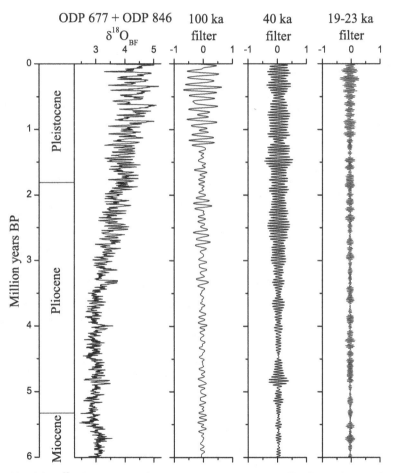

Fig. 6.31 $\delta^{18}O$ from benthic foraminifera at V19-30, ODP Site 677 and ODP Site 846 in the eastern Pacific (Fig. 6.1, Table 1.1) over the last 6 million years. Band-pass filter output plotted at the same horizontal scale as the original time series, designed to extract variability at the orbital periods, reveals the changing response of the climate. Compare the filter output with the orbital parameters over the last 6 million years in Fig. 6.27.

However, early- and mid-Miocene oxygen-isotope records are dominated by obliquity with intervals when 100-ka eccentricity cycles are present but subordinate (Krijgsman *et al.*, 1994b; Beaufort, 1994; Shackleton *et al.*, 1999a; Paul *et al.*, 2000; Zachos *et al.*, 2001). In the Oligocene, obliquity dominates deep-sea stratigraphic records, even at very low latitudes, presumably due to the influence of Antarctic ice sheets on bottom-water production and circulation (Weedon *et al.*, 1997; Zachos *et al.*, 1997; Paul *et al.*, 2000; Zachos *et al.*, 2001). However, the late Eocene was dominated by precession until about 36.5 Ma BP when there was a switch to dominant obliquity variations (Pälike *et al.*, 2001; Sloan and Huber, 2001). Changing dominance of orbital cycles in climate change also appears to have occurred in the Jurassic (Hinnov and Park, 1998, 1999; Weedon *et al.*, 1999).

The causes of the changing dominance of different orbital cycles remain obscure. One possibility is the effect of changing latitudes at which ice sheets and closed lake-basins were formed (Clemens *et al.*, 1996; Hinnov and Park, 1999). In terms of annual changes in insolation, obliquity signals are generally thought to imply a high-latitude origin (e.g. related to ice-sheet growth and cold bottom-water production) whereas precession dominates low-latitude climate change (e.g. Ruddiman and McIntyre, 1984). However, when seasonal temperature thresholds are important to climate change, precession cycles in climate change can be of comparable importance to obliquity cycles, at high and low latitudes (Imbrie *et al.*, 1993b). Short *et al.* (1991) showed that specific land-sea distributions can also modify the seasonal temperature response to the orbitally forced insolation variations. This makes it difficult to identify the latitudinal origin of particular climatic cycles observed in ancient stratigraphic records.

6.9.4b   The 100-ka cycles in the late Pleistocene

As discussed below, one of the best-studied cases of changing climatic responses to orbital forcing concerns the transition from dominantly 40-ka to 100-ka climate cycles in the mid Pleistocene. However, in Mesozoic records there are many cases where 100-ka cycles have been identified (e.g. Herbert and Fischer, 1985; Olsen, 1986; Goldhammer *et al.*, 1990). These Mesozoic cases apparently often result from simple rectification (Section 5.3.1, Crowley *et al.*, 1992; Herbert, 1992).

Imbrie *et al.* (1993a) reviewed the multitude of hypotheses that have been used to account for large-amplitude late-Pleistocene 100-ka cycles that are coherent with eccentricity even though: (a) the 400-ka eccentricity cycle is virtually undetectable; and (b) the eccentricity cycle causes extremely small variations in insolation compared to precession and obliquity. It appears that a non-linear model is required to account for the amplification of eccentricity cycles and one of the simplest and most successful was developed by Imbrie and Imbrie (1980). They generated a system with a single time constant combined with slow rates of ice-sheet growth and fast rates of ice-sheet destruction. This model generates the saw-toothed 100-ka variations in ice volume which are so characteristic of the late Pleistocene.

Muller and Macdonald (1995, 1997a, 1997b) argued that a critical point was that the spectrum of late-Pleistocene oxygen isotopes is dominated by a spectral peak corresponding to cycles with periods of 100 ka. In contrast, a signal related to eccentricity would be expected to have peaks related to cycles with periods of 95 ka, 125 ka and about 400 ka years. They invoked changes in the tilt of the Earth's orbital plane (the ecliptic) which has a period of 100 ka. Their evidence included different spectral signatures from different time series variables from cores from ODP Site 806 in the western Pacific (Muller and MacDonald, 1997a). Climatically this was explained as related to the movement of the Earth's orbit through an interplanetary dust cloud. Some evidence for 100-ka variations in interplanetary dust flux was discovered from ^{3}He measurements (Farley and Patterson, 1995). However, Kortenkamp and Dermott (1998) showed that the interplanetary dust cloud is spread well above and below the limits of the tilt in the ecliptic. Furthermore the changes in interplanetary dust flux are far too small to explain the climate changes proposed by Muller and MacDonald (1995, 1997a). Additionally, the 100-ka variations in dust flux are not explained by the tilt of the ecliptic, but by the variations in eccentricity (Kortenkamp and Dermott, 1998).

Rial (1999) noticed that the period of the 100-ka variations in deep-water oxygen isotopes (inferred global ice volume) was not constant. He argued that the spectrum of deep-water oxygen isotopes has the combination tone peaks (Section 5.2.4) on either side of the main spectral peak near 1/100 ka that are characteristic of frequency modulation (Section 2.2.2). In particular, he argued that the frequency modulation was caused by the 400-ka eccentricity cycle. However, Ridgwell et al. (1999) argued that the spectrum of deep-water oxygen isotopes can be explained by reference to the saw-tooth characteristic emphasized by Imbrie and Imbrie (1980).

A recent important development has been the recognition by Shackleton (2000) that, despite earlier work (Shackleton and Opdyke, 1973), late-Pleistocene deep-water oxygen isotopes are not simply a record of global ice volume. He reasoned that δ^{18}O for air derived from bubbles in the Antarctic Vostok ice core (e.g. Petit et al., 1999) depends mainly on the δ^{18}O of sea water (which depends on global ice volume). After separately allowing for linear orbital forcing in the Vostok and deep-water oxygen isotopes, Shackleton was able to remove the global ice-volume component from the deep-water δ^{18}O. The end result was a residual δ^{18}O signal dominated by 100-ka variations that could only represent changing deep-water temperature.

Shackleton's (2000) paper showed that the late-Pleistocene deep-water temperature is coherent and in phase (Section 4.5) with records of atmospheric CO_2 contents and Antarctic air temperature (as indicated by D/H ratios measured in ice cores). However, the dominantly 100-ka global ice-volume signal lags CO_2, air temperature and deep-water temperature. This is consistent with a model where atmospheric carbon dioxide concentration partly dictates high-latitude air temperatures, deep-water temperatures respond almost instantly to the air temperature variations, but ice-sheet volumes respond with a significant time delay (Shackleton and Pisias, 1985; Shackleton, 2000; Sigman and Boyle, 2000). Thus the Pleistocene 100-ka glacial/interglacial cycles in ice volume apparently relate to the 100-ka variations in CO_2, which now require explanation themselves.

6.10   Chapter overview

■ The background of the climatic spectrum for periods of between 0.7 years and one million years is red, with different spectral slopes for different frequency ranges. In the reconstruction of the climatic spectrum shown in Fig. 6.8, only the annual-, millennial-scale and orbital-tilt cycles produce significant concentrations of power superimposed on the strong red-noise background.

■ Tidal and annual signals can be regarded as virtually periodic. They produce growth bands and sedimentary laminae, tidal bundles or couplets that can be used to generate discrete-signal records. Demonstrating a periodic origin for such stratification in ancient records is important, but can be difficult. For example, strata that are now thought to be tidal laminae in the Precambrian Elatina Formation have been mistaken for varves (i.e. annual). Furthermore, laminae in the modern Santa Monica Basin, that might be interpreted as of annual origin in ancient strata, are apparently only generated during ENSO events.

■ Regular cyclicity in records related to the ENSO, NAO, solar cycles and millennial-scale cycles is relatively much less stationary than that due to Milankovitch cycles.

■ Millennial-scale cyclicity appears to be pervasive globally in modern high-resolution palaeoceanographic records and may have been persistent throughout the Phanerozoic. The origin of millennial-scale cyclicity has yet to be established, but may relate to long-period tidal cycles or solar forcing.

■ Records of Milankovitch cyclicity have proved to be extremely useful for improving the geological time scale in the late Cenozoic via orbital tuning. Older records have provided time scales that have been used for 'interval dating' and in particular for measuring rates of various processes, including biological and geophysical changes.

Appendix – published algorithms for time-series analysis

Some readers will wish to use an existing commercial (e.g. Matlab, Origin, Statistica, SPSS) or internet-derived time-series package. Of the latter the 'SSA-MTM Toolkit' (at http://www.atmos.ucla.edu/tcd/ssa, see Dettinger *et al.*, 1995), 'Analyseries' (at http://www.ngdc.noaa.gov/paleo/softlib.html, see Paillard *et al.*, 1996) and 'Arand' (at http://www.ngdc.noaa.gov/paleo/softlib.html) all provide a variety of time series procedures. Additionally, http://paos.colorado.edu/research/wavelets allows wavelet analysis with the Morlet wavelets. However, the problem with using existing packages is that the speed with which output can be obtained might prevent the unwary from checking which procedures are being used. Always look inside the 'black box' if you can. In particular investigate the documentation to see whether, for example, the data are checked for uneven spacing (including values in the wrong order or multiple values at the same stratigraphic position or time). Will you be able to control the pre-processing of the data (e.g. interpolation, pre-whitening, detrending)? Are appropriate statistical tests available?

Of course the best way to learn how time series methods actually work is to look at the original coding and implement the data processing oneself. This appendix is designed to help this process and includes the algorithms adapted for generating the figures in this book. I have listed references of published FORTRAN algorithms that are in printed form. This has several advantages: (a) the code is independent of the platform (i.e. PC, Mac, Sun) and operating system used, (b) there is always appropriate documentation, (c) there is no risk of viruses and (d) there are no web sites that might become inaccessible/deleted. However, (a) the reader might not be familiar with FORTRAN, (b) data input procedures and graphical output need to be arranged and (c) appropriate statistical tests need to be applied. The first two difficulties are easily surmounted assuming some familiarity with programming. In terms of statistical tests

Table 1. *Table of printed FORTRAN computer algorithms for time-series analysis*

Technique	Source	Notes
Pre-processing		
Interpolation	Press *et al.*, 1992	
Detrending	Bloomfield, 1976	For unevenly spaced data use a
	Press *et al.*, 1992	least-squares fit from Press *et al.*, 1982
Data tapering	Bloomfield, 1976	
	Press *et al.*, 1992	
	Pardo-Igúzquiza *et al.*, 1994	
Fourier transform		
DFT	Davis, 1973	Can use any number of data points.
	Press *et al.*, 1992	Press *et al.* algorithm allows use of unevenly spaced data (implementing the Lomb–Scargle Fourier transform).[1]
FFT	Bloomfield, 1976	All sources require an integer power
	Press *et al.*, 1992	of two data points (e.g. 256, 512,
	Pardo-Igúzquiza *et al.*, 1994	1024..., truncate data or zero-pad).
Spectral estimation		
Direct method	Davis, 1973	Discrete spectral window applied to
	Bloomfield, 1976	periodogram estimates. Bloomfield,
	Press *et al.*, 1992	Press *et al.* and Pardo-Igúzquiza
	Pardo-Igúzquiza *et al.*, 1994	*et al.* require an integer power of two points (truncate or zero-pad).[1]
Multi-taper method	Pardo-Igúzquiza *et al.*, 1994	Requires an integer power of two points (truncate data or zero-pad).[2]
Blackman–Tukey method	Pardo-Igúzquiza *et al.*, 1994	Requires an integer power of two points (truncate data or zero-pad). Employs Bartlett, Parzen or Tukey spectral windows.
Maximum entropy method	Press *et al.*, 1992	Pardo-Igúzquiza *et al.* algorithm
	Pardo-Igúzquiza *et al.*, 1994	requires a power of two points (truncate or zero-pad).
Walsh method	Beauchamp, 1975	Both references require a power of two
	Kanasewich, 1981	points (truncate or zero-pad). NB. The average of Walsh periodograms from phase-shifted records is needed to generate the average Walsh spectrum (invariant with data phase, Section 3.4.6).

Table 1. (*cont.*)

Technique	Source	Notes
Other methods		
Filtering	McClellan *et al.*, 1973 Otnes and Enochson, 1978	McClellan *et al.* provide FIR low-pass, high-pass and band-pass filters (the implementation is explained by Ifeachor and Jervis, 1993). Otnes and Enochson provide the IIR Butterworth low-pass filter (differencing two low-pass filters yields a band-pass filter).
Complex demodulation	Bloomfield, 1976	Output must be low-pass filtered to remove components at the demodulation frequency.
Correlogram	Davis, 1973	Autocorrelation versus lag.
Coherency and phase spectra	Bloomfield, 1976	Requires an integer power of two points (truncate or zero-pad).[1]
SSA	Davis, 1973 Press *et al.*, 1992	Algorithms for determining eigenvalues and eigenvectors. Davis (1973) and Vautard and Ghil (1989) explain reconstruction of principal components.
Wavelets	Press *et al.*, 1992	Uses Daubechies mother wavelets.
Non-linear methods (inc. tests for chaos)	Kantz and Shreiber, 1997	The data must be stationary in the strict sense and have sufficient values (many thousands of points).

[1]Schulz and Stattegger (1997) provide coding in Pascal for spectral and cross-spectral estimation for irregularly spaced data. This is implemented using the Lomb–Scargle Fourier Transform. Spectral smoothing uses the Welch method based on averaging periodograms from overlapping data segments (Section 3.4.2).

[2]Algorithm EIGENJ is required from Davis (1973) for implementing the multi-taper method of Pardo-Igúzquiza *et al.* (1994). This necessitates replacement in EIGENJ of the line 'DIMENSION A(N1,N1), B(N1,N1)' with 'REAL*8 A(N1,N1), B(N1,N1)'.

DFT: discrete Fourier transform; FFT: fast Fourier transform; SSA: singular spectrum analysis.

Bloomfield (1976) provides details for determining resolution bandwidth, as well as confidence intervals and confidence limits for power-, coherency- and phase-spectra. References given here or in the main text discuss statistical tests for the other methods.

Unless otherwise specified, the algorithms listed here (Table 1) require evenly spaced data. In most cases there is a reference to the classic *Numerical Recipes, the Art of Scientific Computing* by Press *et al.* (1992). The advantage of this reference is that the algorithms are also available in FORTRAN 90, Pascal, and C and they can be obtained in electronic form as well as hardcopy. Multi-taper spectral estimation using an algorithm written in C is provided by Lees and Park (1995).

References

All web addresses were live at the time of printing.

Algeo, T.J. (1993). Quantifying stratigraphic completeness: a probabilistic approach using paleomagnetic data. *J. Geol.* **101**: 421–33.

Algeo, T.J. and Woods, A.D. (1994). Microstratigraphy of the Lower Mississippian Sunbury Shale: a record of solar-modulated climatic cyclicity. *Geology* **22**: 795–8.

Allen, J.R.L. (1981). Lower Cretaceous tides revealed by cross-bedding with mud drapes. *Nature* **289**: 579–81.

Allen, M.R. and Smith, L.A. (1996). Monte Carlo SSA: detecting irregular oscillations in the presence of colored noise. *J. Clim.* **9**: 373–404.

Allen, P.A. and Homewood, P. (1984). Evolution and mechanics of a Miocene tidal sandwave. *Sedimentology* **31**: 63–81.

Alley, R.B. and MacAyeal, D.R. (1994). Ice-rafted debris associated with binge/purge oscillations of the Laurentide Ice Sheet. *Paleoceanography* **9**: 503–11.

Alley, R.B., Shuman, C.A., Meese, D.A., Gow, A.J., Taylor, K.C., Cuffey, K.M., Fitzpatrick, J.J., Grootes, P.M., Zielinski, G.A., Ram, M., Spinelli, G. and Elser, B. (1997). Visual-stratigraphic dating of the GISP2 ice core: basis, reproducibility, and application. *J. Geophys. Res.* **102**: 26,367–81.

Alley, R.B., Anandakrishnan, S. and Jung, P. (2001). Stochastic resonance in the North Atlantic. *Paleoceanography* **16**: 190–8.

Anders, M.H., Krueger, S.W. and Sadler, P.M. (1987). A new look at sedimentation rates and the completeness of the stratigraphic record. *J. Geol.* **95**: 1–14.

Anderson, D.M. (2001). Attenuation of millennial-scale events by bioturbation in marine sediments. *Paleoceanography* **16**: 352–7.

Anderson, R.Y. (1961). Solar-terrestrial climatic patterns in varved sediments. *Ann. NY Acad. Sci. USA* **95**: 424–35.

Anderson, R.Y. (1982). A long geoclimatic record from the Permian. *J. Geophys. Res.* **87**: 7285–94.

Anderson, R.Y. (1984). Orbital forcing of evaporite sedimentation. In: *Milankovitch and Climate*, Eds: A. Berger, J. Imbrie, J.D. Hays, G. Kukla, B. Saltzman, NATO ASI Series C126. Reidel, Dordrecht, volume 1, pp. 147–62.

Anderson, R.Y. (1986). The varve microcosm: propagator of cyclic bedding. *Paleoceanography* **1**: 373–82.

Anderson, R.Y. (1992a). Long term changes in the frequency of occurrence of El Niño events. In: *El Niño: Historical and Paleoclimatic Aspects of the Southern Oscillation*, Eds: H.F. Diaz and V. Markgraf. Cambridge University Press, Cambridge, pp. 193–200.

Anderson, R.Y. (1992b). Possible connection between surface winds, solar activity and the Earth's magnetic field. *Nature* **358**: 51–3.

Anderson, R.Y. (1996). Seasonal sedimentation: a framework for reconstructing climatic and environmental change. In: *Palaeoclimatology and Palaeoceanography from Laminated Sediments*, Ed: A.E.S. Kemp. Geological Society Special Publication, No. 116. The Geological Society, London, pp.1–15.

Anderson, R.Y. and Dean, W.E. (1988). Lacustrine varve formation through time. *Palaeogeog. Palaeoclim. Palaeoecol.* **62**: 215–35.

Anderson, R.Y. and Koopmans, L.H. (1963). Harmonic analysis of varve time series. *J. Geophys. Res.* **68**: 877–93.

Andrews, J.T., Jennings, A.E., Kerwin, M., Kirby, M., Manley, W., Miller, G.H., Bond, G. and MacLean, B. (1995). A Heinrich-like event, H-0 (DC-0): source(s) for detrital carbonate in the North Atlantic during the Younger Dryas chronozone. *Paleoceanography* **10**: 943–52.

Andrews, J.T., Barber, D.C. and Jennings, A.E. (1999). Errors in generating time series and in dating events at Late Quaternary millennial (radiocarbon) time-scales: examples from Baffin Bay, NW Labrador Sea, and East Greenland. In: *Mechanisms of Global Climate Change at Millennial Time Scales*, Eds: P.U. Clark, R.S. Webb and L.D. Keigwin. Geophysical Monograph 112. American Geophysical Union, Washington, pp. 23–33.

Appenzeller, C., Stocker, T.F. and Anklin, M. (1998). North Atlantic Oscillation dynamics recorded in Greenland ice cores. *Science* **282**: 446–9.

Archer, A.W. (1996). Reliability of lunar orbital periods extracted from ancient cyclic tidal rhythmites. *Earth Planet. Sci. Lett.* **141**: 1–10.

Archer, A.W. and Johnson, T.W. (1997). Modelling of cyclic tidal rhythmites (Carboniferous of Indiana and Kansas, Precambrian of Utah, USA) as a basis for reconstruction of intertidal positioning and palaeotidal regimes. *Sedimentology* **44**: 991–1010.

Archer, A.W., Kuecher, G.J. and Kvale, E.P. (1995). The role of tidal-velocity asymmetries in the deposition of silty tidal rhythmites (Carboniferous, eastern Interior Coal Basin, U.S.A.). *J. Sed. Res.* **A65**: 408–16.

Arz, H.W., Pätzold, J. and Wefer, G. (1998). Correlated millennial-scale changes in surface hydrography and terrigenous sediment yield inferred from last-glacial marine deposits off Northeastern Brazil. *Quat. Res.* **50**: 157–66.

Baker, P.A., Rigsby, C.A., Sltzer, G.O., Fritz, S.C., Lowenstein, T.K., Bacher, N.P. and Veliz, C. (2001).

Tropical climate changes at millennial and orbital timescales on the Bolivian Altiplano. *Nature* **409**: 698–701.

Barahona, M. and Poon, C-S. (1996). Detection of non-linear dynamics in short, noisy time series. *Nature* **381**: 215–17.

Bard, E. (2001). Paleoceanographic implications of the difference in deep-sea sediment mixing between large and fine particles. *Paleoceanography* **16**: 235–9.

Barlow, L.K., White, J.W.C., Barry, R.G., Rogers, J.C. and Grootes, P.M. (1993). The North Atlantic Oscillation signature in deuterium and deuterium excess signals in the Greenland Ice Sheet Project 2 ice core, 1840–1970. *Geophys. Res. Lett.* **20**: 2901–4.

Barrell, J. (1917). Rhythms and the measurement of geological time. *Bull. Geol. Soc. Am.* **28**: 745–904.

Barry, R.G. and Chorley, R.J. (1998). *Atmosphere Weather and Climate*, seventh edition. Rutledge, London, pp. 1–409.

Beattie, P.D. and Dade, W.D. (1996). Is scaling in turbidite deposition consistent with forcing by earthquakes? *J. Sed. Res.* **66**: 909–15.

Beauchamp, K.G. (1975). *Walsh Functions and their Applications*. Academic Press, London, pp. 1–236.

Beauchamp, K.G. (1984). *Applications of Walsh and Related Functions, with an Introduction to Sequency Theory*. Academic Press, New York, pp. 1–308.

Beaufort, L. (1994). Climatic importance of the modulation of the 100 kyr cycle inferred from 16 m.y. long Miocene records. *Paleoceanography* **9**: 821–34.

Beer, J., Blinov, A., Bonani, G., Finkel, R.C., Hofmann, H.J., Lehmann, B., Oeschger, H., Sigg, A., Schwander, J., Staffelbach, T., Stauffer, B., Suter, M. and Wölfli, W. (1990). Use of [10]Be in polar ice to trace the 11-year cycle of solar activity. *Nature* **347**: 164–6.

Beer, J., Mende, W. and Stellmacher, R. (2000). The role of the sun in climate forcing. *Quat. Sci. Rev.* **19**: 403–15.

Beerbower, J.R. (1964). Cyclothems and cyclic depositional mechanisms in alluvial plain sedimentation. In: Symposium on Cyclic Sedimentation, Ed: D.F. Merriam. *Kansas Geol. Surv. Bull.* **169**: 31–42.

Behl, R.J., and Kennett, J.P. (1996). Brief interstadial events in the Santa Barbara basin, NE Pacific, during the past 60 kyr. *Nature* **379**: 243–6.

Bender., M., Sowers, T., Dickson, M-L., Orchardo, J., Grootes, P., Mayewski, P.A. and Meese, D.A. (1994). Climate correlations between Greenland and Antarctica during the past 100,000 years. *Nature* **372**: 663–6.

Benzi, R., Parisi, G., Sutera, A. and Vulpiani, A. (1982). Stochastic resonance in climatic change. *Tellus* **34**: 10–16.

Berger, A. (1984). Accuracy and frequency stability of the Earth's orbital elements during the Quaternary. In: *Milankovitch and Climate*, Eds: A. Berger, J. Imbrie, J.D. Hays, G. Kukla, B. Saltzman, NATO ASI Series C126. Reidel, Dordrecht, volume 1, pp. 3–39.

Berger, A. (1988). Milankovitch Theory and climate. *Rev. Geophys.* **26**: 624–57.

Berger, A. (1989). The spectral characteristics of pre-Quaternary climate records, an example of the relationship between Astronomical Theory and Geosciences. In: *Climate and Geosciences*, Eds: A. Berger, S. Schneider, and J.C. Duplessy. Kluwer, Dordrecht, pp. 47–76.

Berger, A. and Loutre, M.F. (1994). Astronomical forcing through geological time. In: *Orbital Forcing and Cyclic*

Sequences, Eds: P.L. de Boer and
D.G. Smith. International Association of
Sedimentologists Special Publication 19.
Blackwell, Oxford, pp. 15–24.

Berger, A. and Loutre, M.F. (1997).
Intertropical latitudes and precessional
and half-precessional cycles. *Science* **278**:
1476–8.

Berger, A., Imbrie, J., Hays, J.D., Kukla, G.
and Saltzman, B. (Eds) (1984).
Milankovitch and Climate,
Understanding the Response to
Astronomical Forcing, Eds: A. Berger, J.
Imbrie, J.D. Hays, G. Kukla, B. Saltzman,
NATO ASI Series C126. Reidel,
Dordrecht, 2 volumes, pp. 1–895.

Berger, A., Loutre, M.F. and Dehant, V.
(1989). Influence of the changing lunar
orbit on the astronomical frequencies of
pre-Quaternary insolation patterns.
Paleoceanography **4**: 555–64.

Berger, A., Mélice, J.L. and Van der
Mersch, I. (1990). Evolutive spectral
analysis of sunspot data over the past
300 years. *Philos. Trans. R. Soc. Lond.*
330A: 529–41.

Berger, A., Loutre, M.F. and Mélice, J.L.
(1997). Instability of the astronomical
periods from 1.5 Myr BP to 0.5 Myr BP.
Palaeoclimates **4**: 1–42.

Berggren, W.A., Kent, D.V., Swisher, C.C.
and Aubry, M. (1995). A revised
Cenozoic geochronology and
chronostratigraphy. In: *Geochronology*
Time Scales and Global Stratigraphic
Correlation, Eds: W.A. Berggren,
D.V. Kent, M. Aubry and J. Hardenbol.
Society of Economic Paleontologists and
Mineralogists (SEPM) Special
Publication No. 54, pp. 129–212.

Bezrukov, S.M. and Vodyanoy, I. (1997).
Stochastic resonance in non-dynamical
systems without response thresholds.
Nature **385**: 319–21.

Bianchi, G.G. and McCave, I.N. (1999).
Holocene periodicity in North Atlantic
climate and deep-ocean flow south of
Iceland. *Nature* **397**: 515–17.

Bigg, G.R. (1996). *The Oceans and Climate.*
Cambridge University Press, Cambridge,
pp. 1–266.

Biondi, F., Lange, C.B., Hughes, M.K. and
Berger, W.H. (1997). Inter-decadal
signals during the last millennium
(AD 1117–1992) in the varve record of
Santa Barbara basin, California. *Geophys.*
Res. Lett. **24**: 193–6.

Black, D.E., Peterson, L.C., Overpeck, J.T.,
Kaplan, A., Evans, M.N. and Kashgarian,
M. (1999). Eight centuries of North
Atlantic ocean atmosphere variability.
Science **286**: 1709–13.

Bloomfield, P. (1976). *Fourier Analysis of*
Time Series: an Introduction. Wiley,
London.

Bond, G. and Lotti, R. (1995). Iceberg
discharges into the North Atlantic on
millennial time scales during the last
glaciation. *Science* **267**: 1005–10.

Bond, G.C., Heinrich, H., Broecker, W.S.,
Labeyrie, L., McManus, J.F.,
Andrews, J.T., Huon, S., Jantschik, R.,
Clasen, S., Simet, C., Tedesco, K.,
Klas, M., Bonani, G. and Ivy, S. (1992).
Evidence for massive discharges of
icebergs into the Northern Atlantic Ocean
during the last glacial period. *Nature* **360**:
245–9.

Bond, G., Broecker, W., Johnsen, S.,
McManus, J., Labeyrie, L., Jouzel, J. and
Bonani, G. (1993). Correlations between
climate records from North Atlantic
sediments and Greenland ice. *Nature* **365**:
143–7.

Bond, G., Showers, W., Cheseby, M.,
Lotti, R., Almasi, P., deMenocal, P.,
Priore, P., Cullen, H., Hajdas, I. and
Bonani, G. (1997). A pervasive

millennial-scale cycle in North Atlantic Holocene and glacial climates. *Science* **278**: 1257–66.

Bond, G., Kromer, B., Beer, J., Muscheler, R., Evans, M.N., Showers, W., Hoffmann, S., Lotti-Bond, R., Hajdas, I. and Bonani, G. (2001). Persistent solar influence on North Atlantic climate during the Holocene. *Science* **294**: 2130–6.

Boss, S.K. and Rasmussen, K.A. (1995). Misuse of Fischer plots as sea-level curves. *Geology* **23**: 221–4.

Boygle, J. (1993). The Swedish varve chronology – a review. *Prog. Phys. Geog.* **17**: 1–19.

Bradley, W.H. (1929). The varves and climate of the Green River epoch. *US Geol. Surv. Prof. Pap.* **158**: 87–110.

Briffa, K.R. (2000). Annual climate variability in the Holocene: interpreting the message of ancient trees. *Quat. Sci. Rev.* **19**: 87–105.

Briffa, K.R. and Osborn, T.J. (1999). Seeing the wood from the trees. *Science* **284**: 926–7.

Briffa, K.R., Jones, P.D., Schweingruber, F.H. and Osborn, T.J. (1998a). Influence of volcanic eruptions on Northern hemisphere summer temperature over the past 600 years. *Nature* **393**: 450–5.

Briffa, K.R., Schweingruber, F.H., Jones, P.D., Osborn, T.J., Harris, I.C., Shiyatov, S.G., Vaganov, E.A. and Grudd, H. (1998b). Trees tell of past climates: but are they speaking less clearly today? *Philos. Trans. R. Soc. Lond.* **353B**: 65–73.

Briffa, K.R., Schweingruber, F.H., Jones, P.D., Osborn, T.J., Shiyatov, S.G. and Vaganov, E.A. (1998c). Reduced sensitivity of recent tree-growth to temperature at high northern latitudes. *Nature* **391**: 678–82.

Briffa, K.R., Bartholin, T.S., Eckstein, D., Jones, P.D., Karlén, W., Schweingruber, F.H. and Zetterberg, P. (1990). A 1,400-year tree-ring record of summer temperatures in Fennoscandia. *Nature* **346**: 434–9.

Broecker, W.S. (1997). Thermohaline circulation, the Achilles heel of our climate system: will man-made CO_2 upset the current balance? *Science* **278**: 1582–8.

Broecker, W.S. and Denton, G.H. (1990). The role of ocean-atmosphere reorganizations in glacial cycles. *Geochim. Cosmochim. Acta* **53**: 2464–501.

Broecker, W.S., Bond, G., Klas, M., Bonani, G. and Wolfli, W. (1990). A salt oscillator in the glacial Atlantic? 1, the concept. *Paleoceanography* **5**: 469–77.

Broecker, W.S., Bond, G., Klas, M., Clark, E. and McManus, J.F. (1992). Origin of the northern Atlantic's Heinrich events. *Clim. Dynam.* **6**: 265–73.

Brook, E.J., Sowers, T. and Orchardo, J. (1994). Rapid variations in methane concentration during the past 110,000 years. *Science* **273**: 1087–91.

Buchanan, M. (2000). *Ubiquity. The Science of History . . . or Why the World is Simpler than We Think*. Weidenfield and Nicholson, London, pp. 1–230.

Bull, D., Kemp, A.E.S. and Weedon, G.P. (2000). A 160,000 year old record of El Niño-Southern Oscillation in marine production and coastal run-off from Santa Barbara Basin, California. *Geology* **28**: 1007–10.

Burroughs, W.J. (1992). *Weather Cycles. Real or Imaginary?* Cambridge University Press, Cambridge, pp. 1–207.

Cande, S.C. and Kent, D.V. (1995). Revised calibration of the geomagnetic polarity time scale for the Late Cretaceous and Cenozoic. *J. Geophys. Res.* **100**: 6093–5.

Cane, M.A. (1992). Tropical Pacific ENSO models: ENSO as a mode of the coupled system. In: *Climate System Modelling*, Ed: K.E. Trenberth. Cambridge University Press, Cambridge, pp. 583–614.

Cannariato, K.G., Kennett, J.P. and Behl, R.J. (1999). Biotic response to late Quaternary rapid climate switches in Santa Barbara Basin: ecological and evolutionary implications. *Geology* **27**: 63–6.

Carrs, B.W. and Neidell, N.S. (1966). A geological cyclicity detected by means of polarity coincidence correlation. *Nature* **212**: 136–7.

Chan, M.J., Kvale, E., Archer, A.W. and Sonett, C.P. (1994). Oldest direct evidence of lunar-solar tidal forcing encoded in sedimentary rhythmites, Proterozoic Big Cottonwood Formation, central Utah. *Geology* **22**: 791–4.

Chang, P., Ji, L. and Li, H. (1997). A decadal climate variation in the tropical Atlantic Ocean from thermodynamic air-sea interactions. *Nature* **385**: 516–18.

Christensen, C.J., Gorsline, D.S., Hammond, D.E. and Lund, S.P. (1994). Non-annual laminations and explanation of anoxic basin-floor conditions in Santa Monica Basin, California Borderland, over the past four centuries. *Mar. Geol.* **116**: 399–418.

Cisne, J.L. (1986). Earthquakes recorded stratigraphically on carbonate platforms. *Nature* **323**: 320–2.

Clark, P.U., Webb, R.S. and Keigwin, L.D. (Eds) (1999). *Mechanisms of Global Climate Change at Millennial Time Scales*. Geophysical Monograph 112. American Geophysical Union, Washington, pp. 1–394.

Clark, P.U., Pisias, N.G., Stocker, T.F. and Weaver, A.J. (2002). The role of the thermohaline circulation in abrupt climate change. *Nature* **415**: 863–9.

Cleaveland, M.K., Cook, E.R. and Stahle, D.W. (1992). Secular variability of the Southern Oscillation detected in tree-ring data from Mexico and the southern United States. In: *El Niño: Historical and Paleoclimatic Aspects of the Southern Oscillation*, Eds: H.F. Diaz and V Markgraf. Cambridge University Press, Cambridge, pp. 271–91.

Clemens, S.C. (1999). An astronomical tuning strategy for Pliocene sections: implications for global-scale correlation and phase relationships. *Philos. Trans. R. Soc. Lond.* **357**: 1949–73.

Clemens, S.C. and Prell, W.L. (1990). Late Pleistocene variability of Arabian Sea summer monsoon winds and continental aridity: eolian records from the lithogenic component of deep-sea sediments. *Paleoceanography* **5**: 109–45.

Clemens, S.C., Murray, D.W. and Prell, W.L. (1996). Nonstationary phase of the Plio-Pleistocene Asian Monsoon. *Science* **274**: 943–8.

Clement, A.C., Seager, R. and Cane, M.A. (1999). Orbital controls on the El Niño/Southern Oscillation and the tropical climate. *Paleoceanography* **14**: 441–56.

Cohen, A.L., Layne, G.D., Hart, S.R. and Lobel, P.S. (2001). Kinetic control of skeletal Sr/Ca in a symbiotic coral: implications for the paleotemperature proxy. *Paleoceanography* **16**: 20–6.

Cole, J.E., Dunbar, R.B., McClanahan, T.R. and Muthiga, N.A. (2000). Tropical Pacific forcing of decadal SST variability in the western Indian Ocean over the past two centuries. *Science* **287**: 617–19.

Cook, E.R., D'Arrigo, R.D. and Briffa, K.R. (1998). A reconstruction of the North Atlantic Oscillation using tree-ring chronologies from North America and Europe. *The Holocene* **8**: 9–17.

Corrège, T., Delcroix, T., Récy, J., Beck, W., Caboich, G. and Cornec, F.L. (2000). Evidence for stronger El Niño-Southern Oscillation (ENSO) events in a mid-Holocene massive coral. *Paleoceanography* **15**: 465–70.

Croll, J. (1864). On the physical cause of the change of climate during geological epochs. *Phil. Mag.* **28**: 121–37.

Crowley, K.D., Duchon, C.E. and Rhi, J. (1986). Climate record in varved sediments of the Eocene Green River Formation. *J. Geophys. Res.* **91**: 8637–47.

Crowley, T.J., Kim, K-Y., Mengel, J.G. and Short, D.A. (1992). Modeling 100,000-year climate fluctuations in pre-Pleistocene time series. *Science* **255**: 705–7.

Crusius, J. and Anderson, R.F. (1992). Inconsistencies in accumulation rates of Black Sea sediments inferred from records of laminae and ^{210}Pb. *Paleoceanography* **7**: 215–27.

Cullen, H.M., D'Arrigo, R.D. and Cook, E.R. (2001). Multiproxy reconstructions of the North Atlantic. *Paleoceanography* **16**: 27–39.

Currie, R.G. (1987). Examples and implications of 18.6- and 11-yr terms in world weather records. In: *Climate History, Periodicity and Predictability*, Eds: M. Rampino, J.E. Sanders, W.S. Newman and L.K. Kingsson. Van Nostrand Reinhold, New York, pp. 378–403.

Curry, R.G., McCartney, M.S. and Joyce, T.M. (1998). Oceanic transport of subpolar climate signals to mid-depth subtropical waters. *Nature* **391**: 575–7.

Curry, W.B. and Oppo, D.W. (1997). Synchronous high-frequency oscillations in tropical sea surface temperatures and

North Atlantic deep water production during the last glacial cycle. *Paleoceanography* **12**: 1–14.

Dalfes, H.N., Schneider, S.H. and Thompson, S.L. (1984). Effects of bioturbation on climatic spectra inferred from deep sea cores. In: *Milankovitch and Climate*, Eds: A. Berger, J. Imbrie, J.D. Hays, G. Kula, B. Saltzman, NATO ASI Series C126. Reidel, Dordrecht, volume 1, pp. 481–92.

Dalrymple, R.W. and Makino, Y. (1989). Description and genesis of tidal bedding in the Cobequid Bay – Salmon River estuary, Bay of Fundy, Canada. In: *Sedimentary Facies of Active Plate Margins*, Eds: A. Taira and F. Masuda. Terra, Tokyo, pp. 151–77.

Dansgaard, W., Johnsen, S.J., Clausen, H.B., Dahl-Jensen, D., Gundestrup, N.S., Hammer, C.U., Hvidberg, C.S., Steffensen, J.P., Sveinbjörnsdottir, A.E., Jouzel, J. and Bond, G. (1993). Evidence for general instability of past climate from a 250-kyr ice-core record. *Nature* **364**: 218–20.

D'Arrigo, R.D., Cook, E.R., Jacby, G.C. and Briffa, K.R. (1993). NAO and sea surface temperature signatures in tree-ring records from the North Atlantic sector. *Quat. Sci. Rev.* **12**: 431–40.

Davis, J.C. (1973). *Statistics and Data Analysis in Geology*, first edition. Wiley, London, pp. 1–550.

Davis, J.C. (1986). *Statistics and Data Analysis in Geology*, second edition. Wiley, Chichester, pp. 1–646.

de Boer, P.L. and Smith, D.G. (Eds) (1994). *Orbital Forcing and Cyclic Sequences*. International Association of Sedimentologists Special Publication 19. Blackwell, Oxford, pp. 1–559.

de Boer, P.L. and Wonders, A.A.H. (1984). Astronomically induced rhythmic bedding in Cretaceous pelagic sediments near Moria, Italy. In: *Milankovitch and Climate*, Eds: A. Berger, J. Imbrie, J.D. Hays, G. Kukla, B. Saltzman, NATO ASI Series C126. Reidel, Dordrecht, volume 1, pp. 177–90.

de Boer, P.L., Oost, A.P. and Visser, M.J. (1989). The diurnal inequality of the tide as a parameter for recognizing tidal influences. *J. Sed. Petrol.* **59**: 912–21.

Dehant, V., Loutre, M-F. and Berger, A. (1990). Potential impact of the Northern Hemisphere Quaternary ice sheets on the frequencies of the astroclimatic orbital parameters. *J. Geophys. Res.* **95**: 7573–8.

Dessai, S. and Walter, M.E. (2000). Self-organised criticality and the atmospheric sciences: selected review, new findings and future directions. Contributed paper, NSF Workshop on Extreme Events (http://www.esig.ucar. edu/extremes/papers.html).

Dettinger, M.D., Ghil, M., Strong, C.M., Weibel, W. and Yiou, P. (1995). Software expedites singular-spectrum analysis of noisy time series. *EOS Trans. AGU* **76**: 12–21.

Diaz, H.E. and Markgraf, V. (Eds) (1992). *El Niño: Historical and Paleoclimatic Aspects of the Southern Oscillation.* Cambridge University Press, Cambridge.

Dicke, R.H. (1978). Is there a chronometer hidden deep in the Sun? *Nature* **276**: 676–80.

Dicke, R.H. (1979). Solar luminosity and the sunspot cycle. *Nature* **280**: 24–7.

Dimitrov, B.D., Shangova-Gigoriadi, S. and Grigoriadis, E.D. (1998). Cyclicity in variations of incidence rates for breast cancer in different countries. *Folia Med. (Plovdiv)* **40**: 66–71.

Dokken, T.M. and Jansen, E. (1999). Rapid changes in the mechanism of ocean convection during the last glacial period. *Nature* **401**: 458–61.

Droxler, A.W. and Schlager, W. (1985). Glacial versus interglacial sedimentation rates and oxygen-isotope record in the Bahamas. *Geology* **13**: 799–802.

Duff, P.M.D. and Walton, E.K. (1962). Statistical basis for cyclothems: a quantitative study of the sedimentary succession in the east Pennine coalfield. *Sedimentology* **1**: 235–55.

Duff, P.M.D., Hallam, A. and Walton, E.K. (1967). *Cyclic Sedimentation. Developments in Stratigraphy.* Elsevier, Amsterdam, pp. 1–280.

Dunbar, R.B., Wellington, G.M., Colgan, M.W. and Glynn, P.W. (1994). Eastern Pacific sea surface temperature since 1600AD: the $\delta^{18}O$ record of climatic variability in Galápagos corals. *Paleoceanography* **9**: 291–315.

Dunn, C.E. (1974). Identification of sedimentary cycles through Fourier analysis of geochemical data. *Chem. Geol.* **13**: 217–32.

Duvall, T.L., D'Silva, S., Jefferies, S.M., Harvey, J.W. and Schou, J. (1996). Downflows under sunspots detected by helioseismic tomography. *Nature* **379**: 235–7.

Eddy, J.A. (1976). The Maunder minimum. *Science* **192**: 1189–202.

Egbert, G.D. and Ray, R.D. (2000). Significant dissipation of tidal energy in the deep ocean inferred from satellite altimeter data. *Nature* **405**: 775–8.

Einsele, G. and Seilacher, A. (1991). Distinction of tempestites and turbidites. In: *Cycles and Events in Stratigraphy*, Eds: G. Einsele, W. Ricken and A. Seilacher. Springer, Berlin, pp. 377–82.

Einsele, G., Ricken, W. and Seilacher, A. (1991). Cycles and events in stratigraphy – basic concepts and terms. In: *Cycles and Events in Stratigraphy*, Eds: G. Einsele, W. Ricken and A. Seilacher. Springer, London, pp. 1–19.

Elliot, M., Labeyrie, L., Dokken, T. and Manthe, S. (2000). Coherent patterns of ice rafted debris deposits in the Nordic regions during the last glacial (10–60 ka). *Earth Planet. Sci. Lett.* **194**: 151–63.

Elrick, M. and Hinnov, L.A. (1996). Millennial-scale climate origins for stratification in Cambrian and Devonian deep-water rhythmites, western USA. *Palaeogeog. Palaeoclim. Palaeoecol.* **123**: 353–72.

Elrick, M., Read, J.A. and Coruh, C. (1991). Short-term paleoclimatic fluctuations expressed in lower Mississippian ramp-slope deposits, southwestern Montana. *Geology* **19**: 799–802.

Emery, W.J. and Thomson, R.E. (1997). *Data Analysis Methods in Physical Oceanography*. Elsevier, Amsterdam, pp. 1–634.

Eriksson, K.A. and Simpson, E.L. (2000). Quantifying the oldest tidal record: the 3.2 Ga Moodies Group, Barberton Greenstone Belt, South Africa. *Geology* **28**: 831–4.

Esper, J., Cook, E.R. and Schweingruber, F.H. (2002). Low-frequency signals in long tree-ring chronologies for reconstructing past temperature variability. *Science* **295**: 2250–3.

Evans, J.W. (1972). Tidal growth increments in the cockle *Clinocardium nuttalli*. *Science* **176**: 416–17.

Farley, K.A. and Patterson, D.B. (1995). A 100-kyr periodicity in the flux of extraterrestrial ^{3}He to the seafloor. *Nature* **378**: 600–3.

Faure, G. (1986). *Principles of Isotope Geology*, second edition. Wiley, New York, pp. 1–589.

Fedorov, A.V. and Philander, S.G. (2000). Is El Niño changing? *Science* **288**: 1997–2002.

Finkel, R.C. and Nishiizumi, K. (1997). Beryllium 10 concentrations in the Greenland Ice Sheet Project 2 ice core from 3–40 ka. *J. Geophys. Res.* **102**: 26,699–706.

Fischer, A.G. (1980). Gilbert – bedding rhythms and geochronology. *Spec. Pap. Geol. Soc. Am.* **183**: 93–104.

Fischer, A.G. (1986). Climatic rhythms recorded in strata. *Annu. Rev. Earth Planet. Sci.* **14**: 351–76.

Fischer, A.G. and Bottjer, D.J. (Eds) (1991). Orbital forcing and sedimentary sequences. *J. Sed. Petrol.* **61**: 1063–252.

Fischer, A.G. and Roberts, L.T. (1991). Cyclicity in the Green River Formation (Lacustrine Eocene) of Wyoming. *J. Sed. Petrol.* **61**: 1146–54.

Fischer, A.G. and Schwarzacher, W. (1984). Cretaceous bedding rhythms under orbital control? In: *Milankovitch and Climate*, Eds: A. Berger, J. Imbrie, J.D. Hays, G. Kukla, B. Saltzman, NATO ASI Series C126. Reidel, Dordrecht, volume 1, pp. 163–75.

Fischer, A.G., de Boer, P.L. and Premoli Silva, I. (1990). Cyclostratigraphy. In: *Cretaceous, Resources, Events and Rhythms*, Eds: R.N. Ginsburg and B. Beaudoin. Kluwer, Dordrecht, pp. 139–72.

Fisher, R.A. (1929). Tests of significance in harmonic analysis. *Proc. R. Soc. Series A*, **125**: 54–9.

Forte, A.M. and Mitrovica, J.X.A. (1997). A resonance in the Earth's obliquity and precession over the past 20 Myr driven by mantle convection. *Nature* **390**: 676–9.

Foucault, A., Powichrowski, L. and Prud'Homme, A. (1987). Le côntrole astronomique de la sédimentation turbiditique: exemple du Flysch à Helminthoides des Alpes Ligures (Italie). *C. R. Acad. Sci. Paris* **305**: 1007–11.

Foukal, P. (1990). Solar luminosity variations over timescales of days to the past few solar cycles. *Philos. Trans. R. Soc. Lond.* **330A**: 591–9.

Francois, R. and Bacon, M. (1994). Heinrich events in the North Atlantic: radiochemical evidence. *Deep-Sea Res.* **41**: 315–34.

Friis-Christensen, E. and Lassen, K. (1991). Length of the solar cycle: an indicator of solar activity closely associated with climate. *Science* **254**: 698–700.

Fröhlich, C. and Lean, J. (1998). The Sun's total irradiance: cycles, trends and related climate change uncertainties since 1976. *Geophys. Res. Lett.* **25**: 4377–80.

Fronval, T., Jansen, E., Bloemendal, J. and Johnsen, S. (1995). Oceanic evidence for coherent fluctuations in Fennoscandian and Laurentide ice sheets on millennium timescales. *Nature* **374**: 443–6.

Gagan, M.K., Ayliffe, L.K., Beck, J.W., Cole, J.E., Druffel, E.R.M., Dunbar, R.B. and Schrag, D.P. (2000). New views of tropical paleoclimates from corals. *Quat. Sci. Rev.* **19**: 45–64.

Gallois, R.W. (2000). The stratigraphy of the Kimmeridge Clay (Upper Jurassic) in the RGGE Project boreholes at Swanworth Quarry and Metherhills, south Dorset. *Proc. Geol. Assoc.* **111**: 265–80.

Ganopolski, A. and Rahmstorf, S. (2001). Rapid changes of global climate simulated in a coupled climate model. *Nature* **409**: 153–8.

Ganopolski, A. and Rahmstorf, S. (2002). Abrupt glacial climate changes due to stochastic resonance. *Phys. Rev. Lett.* **88**: Article 038501.

Gershenfeld, N., Schoner, B. and Metois, E. (1999). Cluster-weighted modelling for time-series analysis. *Nature* **397**: 329–32.

Gilbert, G.K. (1895). Sedimentary measurement of geologic time. *J. Geol.* **3**: 351–76.

Gilliland, R.L. (1981). Solar radius variations over the past 265 years. *Astrophys. J.* **248**: 1144–55.

Gipp, M.R. (2001). Interpretation of climate dynamics from phase space portraits: is the climate system strange or just different? *Paleoceanography* **16**: 335–51.

Glatzmaier, G.A., Coe, R.S., Hongre, L. and Roberts, P.H. (1999). The role of the Earth's mantle in controlling the frequency of geomagnetic reversals. *Nature* **401**: 885–90.

Goldhammer, R.K., Dunn, P.A. and Hardie, L.A. (1990). Depositional cycles, composite sea-level changes, cycle stacking patterns and the hierarchy of stratigraphic forcing: examples from Alpine Triassic platform carbonates. *Geol. Soc. Am. Bull.* **102**: 535–62.

Gorsline, D.S., Nava-Sanchez, E. and Murillo de Nava, J. (1996). A survey of occurrences of Holocene laminated sediments in California borderland basins: products of a variety of depositional processes. In: *Palaeoclimatology and Palaeoceanography from Laminated Sediments*, Ed: A.E.S. Kemp. Geological Society Special Publication No. 116. The Geological Society, London, pp. 93–110.

Gough, D. (2000). News from the solar interior. *Science* **287**: 2434–5.

Gradstein, F.M., Agterberg, F.P., Ogg, J.G., Hardenbol, J., Van Veen, P., Thierry, J. and Huang, Z. (1994). A Mesozoic Time Scale. *J Geophys. Res.* **99**: 24,051–74.

Grassberger, P. (1986). Do climatic attractors exist? *Nature* **323**: 609–12.

Greenland Ice-core Project Members (1993). Climate instability during the last interglacial period recorded in the GRIP ice core. *Nature* **364**: 203–7.

Grootes, P.M. and Stuiver, M. (1997). Oxygen 18/16 variability in Greenland snow and ice with 10^{-3} to 10^5-year time resolution. *J. Geophys. Res.* **102**: 26,455–470.

Gu, D. and Philander, S.G.H. (1997). Interdecadal climate fluctuations that depend on exchanges between the tropics and extratropics. *Science* **275**: 805–7.

Guyodo, Y. and Valet, J-P. (1999). Global changes in intensity of the Earth's magnetic field during the past 800 kyr. *Nature* **399**: 249–52.

Gwiazda, R.H., Hemming, S.R. and Broecker, W.S. (1996). Provenance of icebergs during Heinrich event 3 and the contrast to their sources during other Heinrich episodes. *Paleoceanography* **11**: 371–8.

Haak, A.B. and Schlager, W. (1989). Compositional variations in calciturbidites due to sea-level fluctuations, late Quaternary, Bahamas. *Geol. Runds.* **78**: 477–86.

Hagadorn, J.W. (1996). Laminated sediments of Santa Monica Basin, California continental borderland. In: *Palaeoclimatology and Palaeoceanography from Laminated Sediments*, Ed: A.E.S. Kemp. Geological Society Special Publication No. 116. The Geological Society, London, pp. 111–20.

Hagelberg, T.K. and Pisias, N.G. (1990). Nonlinear response of Pliocene climate to orbital forcing: evidence from the eastern equatorial Pacific. *Paleoceanography* **5**: 595–617.

Hagelberg, T.K., Pisias, N.G. and Elgar, S. (1991). Linear and nonlinear couplings between orbital forcing and the marine

δ^{18}O record during the late Neogene. *Paleoceanography* **6**: 729–46.

Hagelberg, T.K., Bond, G. and deMenocal, P. (1994). Milankovitch band forcing of sub-Milankovitch climate variability during the Pleistocene. *Paleoceanography* **9**: 545–58.

Haigh, J.D. (1996). The impact of solar variability on climate. *Science* **272**: 981–4.

Halfman, J.D. and Johnson, T.C. (1988). High resolution record of cyclic climatic change during the past 4 ka from Lake Turkana, Kenya. *Geology* **16**: 496–500.

Hall, I.R., McCave, I.N., Shackleton, N.J., Weedon, G.P. and Harris, S.E. (2001). Glacial intensification of deep Pacific inflow and ventilation. *Nature* **412**: 809–12.

Hallam, A. (1964). Origin of the limestone-shale rhythms in the Blue Lias of England: a composite theory. *J. Geol.* **72**: 157–68.

Hammer, C., Mayewski, P.A., Peel, D. and Stuiver, M. (Eds) (1997). GISP2 and GRIP results. *J. Geophys. Res.* **102** part C12.

Hansen, D.V. and Bezdek, H.F. (1996). On the nature of decadal anomalies in North Atlantic sea surface temperature. *J. Geophys. Res.* **101**: 9749–58.

Hartmann, W.M. (1999). How we localize sound. *Phys. Today* **52**: 24–9.

Hays, J.D., Imbrie, I. and Shackleton, N.J. (1976). Variations in the Earth's orbit: pacemaker of the ice ages. *Science* **194**: 1121–32.

Heinrich, H. (1988). Origin and consequences of cyclic ice rafting in the northeast Atlantic Ocean during the past 130,000 years. *Quat. Res.* **29**: 143–52.

Hendy, I.L. and Kennett, J.P. (1999). Latest Quaternary North Pacific surface-water responses imply atmosphere-driven climate instability. *Geology* **27**: 291–4.

Herbert, T.D. (1992). Paleomagnetic calibration of Milankovitch cyclicity in Lower Cretaceous sediments. *Earth Planet. Sci. Lett.* **112**: 15–28.

Herbert, T.D. (1993). Differential compaction in lithified deep-sea sediments is not evidence for "diagenetic unmixing". *Sediment. Geol.* **84**: 115–22.

Herbert, T.D. (1994). Reading orbital signals distorted by sedimentation: models and examples. In: *Orbital Forcing and Cyclic Sequences*, Eds: P.L. de Boer and D.G. Smith. International Association of Sedimentologists Special Publication 19. Blackwell, Oxford, pp. 483–507.

Herbert, T.D. and Fischer, A.G. (1985). Milankovitch climatic origin of mid-Cretaceous black shale rhythms in central Italy. *Nature* **321**: 739–43.

Herschel, J.F.W. (1832). On the astronomical causes which may influence geological phenomena. *Trans. Geol. Soc. 2nd Ser.* **3**: 393–9.

Hibler, W.D. and Johnsen, S.J. (1979). The 20-yr cycle in Greenland ice core records. *Nature* **280**: 481–3.

Hilgen, F.J. (1991). Astronomical calibration of Gauss to Matuyama sapropels in the Mediterranean and implications for the geomagnetic polarity timescale. *Earth Planet. Sci. Lett.* **104**: 226–44.

Hilgen, F.J. and Langereis, G.C. (1988). The age of the Miocene-Pliocene boundary in the Capo Rossello area (Sicily). *Earth Planet. Sci. Lett.* **91**: 214–22.

Hilgen, F.J. and Langereis, C.G. (1989). Periodicities of CaCO$_3$ cycles in the Pliocene of Sicily: discrepancies with the quasi-periods of the Earth's orbital cycles? *Terra Nova*, **1**: 409–15.

Hilgen, F.J., Krijgsman, W., Langereis, C.G., Lourens, L.J., Santarelli, A. and Zachariasse, W.J. (1995). Extending the astronomical (polarity) time scale into the Miocene. *Earth Planet. Sci. Lett.* **136**: 495–510.

Hilgen, F.J., Abdul Aziz, W., Krijgsman, W., Langereis, C.G., Lourens, L.J., Meulenkamp, J.E., Raffi, I., Steenbrink, J., Turco, E., van Vught, N., Wijbrans, J.R. and Zachariasse, W.J. (1999). Present status of the astronomical (polarity) time-scale for the Mediterranean Late Neogene. *Philos. Trans. R. Soc.* **357**: 1931–47.

Hilgen, F., Schwarzacher, W. and Strasser, A. (2001). Concepts and definitions in cyclostratigraphy – Second Report of the cyclostratigraphy working group. (International Subcommission on Stratigraphic Nomenclature, IUGS Commission of Stratigraphy). Unpublished, 8p.

Hinnov, L.A. (2000). New perspectives on orbitally-forced stratigraphy. *Annu. Rev. Earth Planet. Sci.* **28**: 419–75.

Hinnov, L.A. and Goldhammer, R.K. (1991). Spectral analysis of the Middle Triassic Latemar Limestone. *J. Sed. Petrol.* **61**: 1173–93.

Hinnov, L.A. and Park, J. (1998). Detection of astronomical cycles in the stratigraphic record by frequency modulation (FM) analysis. *J. Sed. Res.* **68**: 524–39.

Hinnov, L.A. and Park, J. (1999). Strategies for assessing Early-Middle (Pliensbachian-Aalenian) Jurassic cyclochronologies. *Philos. Trans. R. Soc. Lond.* **357**: 1831–59.

Holmgren, K., Karlén, W., Lauritzen, S.E., Lee-Thorp, J.A., Partridge, T.C., Piketh, S., Repinski, P., Stevenson, C., Svanered, O. and Tyson, P.D. (1999). A 3000-year high-resolution stalagmite-based record of palaeoclimate for northeastern South Africa. *The Holocene* **9**: 295–309.

House, M.R. (1985). A new approach to an absolute timescale from measurements of orbital cycles and sedimentary microrhythms. *Nature* **315**: 721–5.

House, M.R. and Farrow, G.E. (1968). Daily growth banding in the shell of the cockle, *Cardium edule*. *Nature* **219**: 1384–6.

House, M.R. and Gale, A.S. (Eds) (1995). *Orbital Forcing Timescales and Cyclostratigraphy*. Geological Society Special Publication No. 85. The Geological Society, London, pp. 51–66.

Howe, R., Christensen-Dalsgaard, J., Hill, F., Komm, R.W., Larsen, R.M., Schou, J., Thompson, M.J. and Toomre, J. (2000). Dynamic variations at the base of the solar convective zone. *Science* **287**: 2456–60.

Hoyt, V. and Schatten, K.H. (1997). *The Role of the Sun in Climate Change*. Oxford University Press, Oxford, pp. 1–279.

Hubbard, B.B. (1996). *The World According to Wavelets. The Story of a Mathematical Technique in the Making*. A.K. Peters, Wellesley, Massachusetts, pp. 1–264.

Hulme, M. and Barrow, E. (Eds) (1997). *Climates of the British Isles*. Routledge, London, pp. 1–454.

Hunt, A.G. and Malin, P.E. (1998). Possible triggering of Heinrich events by ice-load-induced earthquakes. *Nature* **393**: 155–8.

Hurrell, J.W. (1995). Decadal trends in the North Atlantic Oscillation: regional temperatures and precipitation. *Science* **269**: 676–9.

Hydrographic Office (1996). *Admiralty Tide Tables 1997, Volume. 1, United Kingdom and Ireland including European Channel Ports*. Hydrographic Office (UK).

Ifeachor, E.C. and Jervis, B.W. (1993). *Digital Signal Processing. A Practical Approach*. Addison-Wesley, Harlow, pp. 1–760.

Imbrie, J. and Imbrie, J.Z. (1980). Modelling the climate response to orbital variations. *Science* **207**: 943–53.

Imbrie, J. and Imbrie, K.P. (1979). *Ice Ages. Solving the Mystery*. Harvard University Press, Cambridge, Massachusetts, pp. 1–224.

Imbrie, J., Hays, J.D., Martinson, D.G., McIntyre, A.C., Mix, A.C., Morley, J.J., Pisias, N.G., Prell, W.L. and Shackleton, N.J. (1984). The orbital theory of Pleistocene climate: support from a revised chronology of the marine $\delta^{18}O$ record. In: *Milankovitch and Climate*, Eds: A. Berger, J. Imbrie, J.D. Hays, G. Kukla, B. Saltzman, NATO ASI Series C126. Reidel, Dordrecht, volume 1, pp. 269–305.

Imbrie, J., Berger, A., Boyle, E.A., Clemens, S.C., Duffy, A., Howard, W.R., Kukla, G., Kutzbach, J., Martinson, D.G., McIntyre, A., Mix, A.C., Molfino, B., Morley, J.J., Peterson, L.C., Pisias, N.G., Prell, W.L., Raymo, M.E., Shackleton, N.J. and Toggweiler, J.R. (1992). On the structure and origin of major glaciation cycles, 1: Linear responses to Milankovitch forcing. *Paleoceanography* **7**: 701–38.

Imbrie, J., Berger, A., Boyle, E.A., Clemens, S.C., Duffy, A., Howard, W.R., Kukla, G., Kutzbach, J., Martinson, D.G., McIntyre, A., Mix, A.C., Molfino, B., Morley, J.J., Peterson, L.C., Pisias, N.G., Prell, W.L., Raymo, M.E., Shackleton, N.J. and Toggweiler, J.R. (1993a). On the structure and origin of the major glaciation cycles, 2: The 100,000-year cycle. *Paleoceanography* **8**: 699–735.

Imbrie, J., Berger, A. and Shackleton, N.J. (1993b). Role of orbital forcing: a two-million-year perspective. In: *Global Changes in the Perspective of the Past*, Eds: J.A. Eddy and H. Oeschger. Wiley, Chichester, pp. 263–77.

Ito, M., Nishikawa, T. and Sugimoto, H. (1999). Tectonic control of high-frequency depositional sequences with durations shorter than Milankovitch cyclicity: an example from the Pleistocene paleo-Tokyo Bay, Japan. *Geology* **27**: 763–6.

James, I.N. and James, P.M. (1989). Ultra-low-frequency variability in a simple atmospheric circulation model. *Nature* **342**: 53–5.

Jenkins, G.M. and Watts, D.G. (1969). *Spectral Analysis and its Applications.* Holden-Day, London.

Jin, F-F., Neelin, J.D. and Ghil, M. (1994). El Niño on the Devil's staircase: annual subharmonic steps to chaos. *Science* **264**: 70–2.

Johnsen, S.J., Clausen, H.B., Dansgaard, W., Gundestrup, N.S., Hammer, C.U., Andersen, U., Andersen, K.K., Hvidberg, C.S., Dahl-Jensen, D., Steffensen, J.P., Shoji, H., Sveinbjörnsdottir, Á.E., White, J., Jouzel, J. and Fisher, D. (1997). The $\delta^{18}O$ record along the Greenland Ice Core Project deep ice core and the problem of possible Eemian climatic instability. *J. Geophys. Res.* **102**: 26,397–410.

Jones, C.E., Jenkyns, H.C. and Hesselbo, S.B. (1994). Strontium isotopes in early Jurassic seawater. *Geochim. Cosmochim. Acta* **58**: 3061–74.

Kanasewich, E.R. (1981). *Time Sequence Analysis in Geophysics.* University of Alberta Press, Alberta.

Kantz, H. and Schreiber, T. (1997). *Nonlinear Time Series Analysis.* Cambridge University Press, Cambridge, pp. 1–304.

Keeling, C.D. and Whorf, T.P. (2000). The 1,800-oceanic tidal cycle: a possible cause of rapid climate change. *Proc. Natl. Acad. Sci. USA* **97**: 3814–19.

Keigwin, L.D. and Lehmann, S.J. (1994). Deep circulation change linked to HEINRICH event 1 and Younger Dryas in a mid depth North Atlantic core. *Paleoceanography* **9**: 185–94.

Kemp, A.E.S. (Ed.) (1996). *Palaeoclimatology and Palaeoceanography from Laminated Sediments.* Geological Society Publication No. 116. The Geological Society, London.

Kemp, A.E.S. and Baldauf, J.G. (1993). Vast Neogene diatom mat deposits from the eastern equatorial Pacific Ocean. *Nature* **362**: 141–4.

Kemp, A.E.S., Baldauf, J.G. and Pearce, R.B. (1995). Origins and paleoceanographic significance of laminated diatom ooze from the eastern equatorial Pacific. *Proc. Ocean Drill. Prog. Sci. Res.* **138**: 641–5.

Kennedy, J.A. and Brassell, S.C. (1992). Molecular records of twentieth-century El Niño events in laminated sediments from the Santa Barbara basin. *Nature* **357**: 62–4.

Kennett, J.P. and Ingram, B.L. (1995). A 20,000-year record of ocean circulation and climate change from the Santa Barbara Basin. *Nature* **377**: 510–14.

Kent, D.V. (1999). Orbital tuning of geomagnetic polarity time-scales. *Philos. Trans. R. Soc. Lond.* **357**: 1995–2007.

Kerr, R.A. (1999). Link between sunspots, stratosphere buoyed [sic]. *Science* **284**: 234–5.

King, T. (1996). Quantifying nonlinearity and geometry in time series of climate. *Quat. Sci. Rev.* **15**: 247–66.

Kirchner, J.W. (2002). Evolutionary speed limits inferred from the fossil record. *Nature* **415**: 65–8.

Kominz, M.A. (1996). Whither cyclostratigraphy? Testing the gamma method on upper Pleistocene deep-sea sediments, North Atlantic Deep Sea Drilling Project site 609. *Paleoceanography* **11**: 481–504.

Kominz, M.A. and Bond, G. (1990). A new method of testing periodicity in cyclic sediments – application to the Newark Supergroup. *Earth Planet. Sci. Lett.* **98**: 233–44.

Kominz, M.A., Beavan, J., Bond, G.C. and McManus, J. (1991). Are cyclic sediments periodic? Gamma analysis and spectral analysis of Newark Supergroup lacustrine strata. In: *Sedimentary Modeling: Computer Simulations and Methods for Improved Parameter Definition*, Eds: K. Franseen, W.L. Watney, C.G.St.C. Kendall and W. Ross. Bull. Kansas Geol. Surv. **233**: 319–34.

Kortenkamp, S.J. and Dermott, S.F. (1998). A 100,000-year periodicity in the accretion rate of interplanetary dust. *Science* **280**: 874–6.

Kotilainen, A.T. and Shackleton, N.J. (1995). Rapid climate variability in the North Pacific Ocean during the past 95,000 years. *Nature* **377**: 323–6.

Krijgsman, W., Hilgen, F.J., Langereis, C.G. and Zachariasse, W.J. (1994a). The age of the Tortonian–Messinian boundary. *Earth Planet. Sci. Lett.* **121**: 533–47.

Krijgsman, W., Langereis, C.G., Daams, R. and Van der Meulen, A.J. (1994b). Magnetostratigraphic dating of the Middle Miocene climate change in the continental deposits of the Aragonian type area in the Calatayud-Teruel basin (central Spain). *Earth Planet. Sci. Lett.* **128**: 513–26.

Krijgsman, W., Hilgen, F.J., Langereis, C.G., Santarelli, A. and Zachariasse, W.J. (1995). Late Miocene magnetostratigraphy, biostratigraphy and cyclostratigraphy in the Mediterranean. *Earth Planet. Sci. Lett.* **136**: 475–94.

Kumar, K.K., Rajagopalan, B. and Cane, M.A. (1999). On the weakening relationship between the Indian Monsoon and ENSO. *Science* **284**: 2156–9.

Kumar, P. and Foufoula-Georgiou, E. (1994). Wavelet analysis in geophysics: an introduction. In: *Wavelets in Geophysics*. Academic Press, London, pp. 1–43.

Kvale, E.P., Archer, A.W. and Johnson, H.R. (1989). Daily, monthly, and yearly tidal cycles within laminated siltstone of the Mansfield Formation (Pennsylvanian) of Indiana. *Geology* **17**: 365–8.

Kvale, E.P., Johnson, H.W., Sonett, C.P., Archer, A.W. and Zawistoski, A. (1999). Calculating the lunar retreat rates using tidal rhythmites. *J. Sed. Res.* **69**: 1154–68.

Labitzke, K. and van Loon, H. (1990). Associations between 11-year solar cycle, the quasi-biennial oscillation and the atmosphere: a summary of recent work. *Philos. Trans. R. Soc. Lond.* **330A**: 577–89.

Lang, W.D., Spath, L.F., Cox, L.R. and Muir-Wood, H.M. (1928). The Belemnite Marls of Charmouth, a series in the Lower Lias of the Dorset Coast. *Quart. J. Geol. Soc.* **84**: 179–257.

Laskar, J. (1989). A numerical experiment on the chaotic behaviour of the solar system. *Nature* **338**: 237–8.

Laskar, J. (1999). The limits of Earth orbital calculations for geological time-scale use. *Philos. Trans. R. Soc. Lond.* **357**: 1735–59.

Laskar, J., Joutel, F. and Robutel, P. (1993a). Stabilization of the Earth's obliquity by the Moon. *Nature* **361**: 615–17.

Laskar, J., Joutel, F. and Boudin, F. (1993b). Orbital, precessional, and insolation quantities for the Earth from −20Myr to +10Myr. *Astron. Astrophys.* **270**: 522–33.

Latif, M. and Barnett, T.P. (1994). Causes of decadal climate variability over the North Pacific and North America. *Science* **266**: 634–7.

Lau, K-M. and Sheu, P.J. (1988). Annual cycle, Quasi-Biennial Oscillation, and Southern Oscillation in global precipitation. *J. Geophys. Res.* **93**: 10,975–88.

Lees, J.M. and Park, J. (1995). Multi-taper spectral analysis: a stand-alone C-subroutine. *Comput. Geosci.* **21**: 199–236.

Le Treut, H. and Ghil, M. (1983). Orbital forcing, climatic interactions, and glaciation cycles. *J. Geophys. Res.* **88**: 5167–90.

Liu, P.C. (1994). Wavelet spectrum analysis and ocean wind waves. In: *Wavelets in Geophysics*. Academic Press, London, pp. 151–66.

Lockwood, M., Stamper, R. and Wild, M.N. (1999). A doubling of the Sun's coronal magnetic field during the past 100 years. *Nature* **399**: 437–9.

Lourens, J.L., Antonarakou, A., Hilgen, F.J., Van Hoof, A.A.M., Vergnaud-Grazzini, C. and Zachariasse, W.J. (1996). Evaluation of the Plio-Pleistocene astronomical timescale. *Paleoceanography* **11**: 391–413.

Lourens, L.J., Wehausen, R. and Brumsack, H.J. (2001). Geological constraints on tidal dissipation and dynamical ellipticity of the Earth over the past three million years. *Nature* **409**: 1029–33.

Lowrie, W. (1997). *Fundamentals of Geophysics*. Cambridge University Press, Cambridge, pp. 1–354.

Lund, D.C. and Mix, A.C. (1998). Millennial-scale deep water oscillations: reflections of the North Atlantic in the deep Pacific from 10 to 60 ka. *Paleoceanography* **13**: 10–19.

Lyell, C. (1830). *Principles of Geology*, Volume 1. John Murray, London.

Lyon, J.G. (2000). The solar wind-magnetosphere-ionosphere system. *Science* **288**: 1987–91.

MacAyeal, D.R. (1993a). A low-order model of the Heinrich event cycle. *Paleoceanography* **8**: 767–73.

MacAyeal, D.R. (1993b). Binge/Purge oscillations of the Laurentide Ice Sheet as a cause of the North Atlantic's Heinrich events. *Paleoceanography* **8**: 775–84.

Mallat, S. (1998). *A Wavelet Tour of Signal Processing*. Academic Press, London, pp. 1–577.

Mann, M.E. and Bradley, R.S. (1999). Northern hemisphere temperatures during the past millennium: inferences, uncertainties and limitations. *Geophys. Res. Lett.* **26**: 759–62.

Mann, M.E. and Lees, J. (1996). Robust estimation of background noise and signal detection in climatic time series. *Clim. Change* **33**: 409–45.

Mann, M.E., Bradley, R.S. and Hughes, M.K. (1998). Global-scale temperature patterns and climate forcing over the past six centuries. *Nature* **392**: 779–87.

Markson, R. and Muir, M. (1980). Solar wind control of the Earth's electric field. *Science* **208**: 979–90.

Martin, E.E., Shackleton, N.J., Zachos, N.J. and Flower, B.P. (1999). Orbitally-tuned Sr isotope chemostratigraphy for the late middle to late Miocene. *Paleoceanography* **14**: 74–83.

Martinson, D.G., Menke, W. and Stoffa, P. (1982). An inverse approach to signal correlation. *J. Geophys. Res.* **87**: 4807–18.

Matsuoka, J., Kano, A., Oba, T., Watanabe, T., Sakai, S. and Seto, K. (2001). Seasonal variation of stable isotopic composition recorded in a laminated tufa, SW Japan. *Earth Planet. Sci. Lett.* **192**: 31–44.

Mayewski, P.A., Meeker, L.D., Twickler, M.S., Whitlow, S., Yang, Q., Lyons, W.B. and Prentice, M. (1997). Major features and forcing of high-latitude northern hemispheric atmospheric circulation using a 110,000-year-long glaciochemical series. *J. Geophys. Res.* **102**: 26,345–66.

McClellan, J.H., Parks, T.W. and Rabiner, L.R. (1973). A computer program for designing optimum FIR linear phase digital filters. *IEEE Trans. Audio Electroacoustics* **21**: 506–26.

McIntyre, A. and Molfino, B. (1996). Forcing of Atlantic equatorial and subpolar millennial cycles by precession. *Science* **274**: 1867–70.

McManus, J.F., Bond, G.C., Broecker, W.S., Johnsen, S., Labeyrie, L. and Higgins, S. (1994). High-resolution climate records from the North Atlantic during the last interglacial. *Nature* **371**: 326–9.

McManus, J.F., Bond, G.C., Broecker, W.S., Fleisher, M.Q. and Higgins, S.M. (1998). Radiometrically determined fluxes in the sub-polar North Atlantic during the last 140,000 years. *Earth Planet. Sci. Lett.* **135**: 29–43.

McManus, J.F., Oppo, D.W. and Cullen, J.L. (1999). A 0.5 million-year record of millennial-scale climate variability in the North Atlantic. *Science* **283**: 971–5.

McPhaden, M.J. and Yu, X. (1999). Equatorial waves and the 1997–98 El Niño. *Geophys. Res. Lett.* **26**: 2961–4.

Medio, A. (1992). *Chaotic Dynamics. Theory and Applications to Economics.* Cambridge University Press, Cambridge, pp. 1–344.

Meeker, L.D., Mayewski, P.A., Grootes, P.M., Alley, R.B. and Bond, G.C. (2001). Comment: "On sharp spectral lines in the climate record and the millennial peak" by C. Wunsch. *Paleoceanography* **16**: 544–7.

Meko, D.M. (1992). Spectral properties of tree-ring data in the United States Southwest as related to El Niño/Southern Oscillation. In: *El Niño: Historical and Paleoclimatic Aspects of the Southern Oscillation*, Ed: H.F. Diaz and V. Markgraf. Cambridge University Press, Cambridge, pp. 227–41.

Melnyk, D.H., Smith, D.G. and Amiri-Garroussi, K. (1994). Filtering and frequency mapping as tools in subsurface cyclostratigraphy, with examples from the Wessex Basin, UK. In: *Orbital Forcing and Cyclic Sequences*, Eds: P.L. de Boer and D.G. Smith. International Association of Sedimentologists Special Publication 19. Blackwell, Oxford, pp. 35–46.

Meyers, S.R., Sageman, B.R. and Hinnov, L.A. (2001). Integrated quantitative stratigraphy of the Cenomanian-Turonian Bridge Creek Limestone Member using evolutive harmonic analysis and stratigraphic modeling. *J. Sed. Res.* **71**: 628–44.

Middleton, G.V., Plotnick, R.E. and Rubin, D.M. (1995). *Nonlinear Dynamics and Fractals. New Numerical Techniques for Sedimentary Data*. Society of Economic Paleontologists and Mineralogists (SEPM) Short Course No. 36. SEPM, Tulsa, Oklahoma, pp. 1–174.

Milankovitch, M. (1941). Canon of insolation in the Ice-Age problem. [English translation by Israel Program for scientific translation, Jerusalem 1969.] *R. Serbian Acad. Spec. Publ.* 132.

Miller, D.J. and Eriksson, K.A. (1997). Late Mississippian prodeltaic rhythmites in the Appalachian Basin: a hierarchical record of tidal and climatic periodicities. *J. Sed. Res.* **67**: 653–60.

Miller, K.G., Feigenson, M.D., Wright, J.D. and Clement, B.M. (1991). Miocene isotope reference section, Deep Sea Drilling Project Site 608: an evaluation of isotope and biostratigraphic resolution. *Paleoceanography* **6**: 33–52.

Mitchell, J.M. (1976). An overview of climatic variability and its causal mechanisms. *Quat. Res.* **6**: 481–93.

Mitrovica, J.X., Forte, A.M. and Pan, R. (1997). Glaciation-induced variations in the Earth's precession frequency, obliquity and insolation over the last 2.6 Ma. *Geophys. J. Int.* **128**: 270–84.

Molinie, A.J., Ogg, J.G. and Ocean Drilling Program Leg 129 Scientific Party (1990). Sedimentation rate curves and discontinuities from sliding-window spectral analysis of logs. *Log Analyst*, November–December, pp. 370–4.

Morgans-Bell, H.S., Coe, A.L., Hesselbo, S.P., Jenkyns, H.C., Weedon, G.P., Marshall, J.E.A. and Williams, C.J. (2001). Integrated stratigraphy of the Kimmeridge Clay Formation (Upper Jurassic) based on exposures and boreholes in South Dorset, UK. *Geol. Mag.* **138**: 511–39.

Morley, C.K., Vanhauwaert, P. and De Batist M. (2000). Evidence for high-frequency cyclic fault activity from high-resolution seismic reflection survey, Rukwa Rift, Tanzania. *J. Geol. Soc. Lond.* **157**: 983–94.

Mudelsee, M. and Stattegger, K. (1994). Plio/Pleistocene climate modeling based on oxygen isotope time series from deep-sea sediment cores: the Grassberger-Procaccia algorithm and chaotic systems. *Math. Geol.* **26**: 799–815.

Muller, R.A. and Macdonald, G.J. (1995). Glacial cycles and orbital inclination. *Nature* **377**: 107–8.

Muller, R.A. and Macdonald, G.J. (1997a). Simultaneous presence of orbital inclination and eccentricity in proxy climate records from Ocean Drilling Program Site 806. *Geology* **25**: 3–6.

Muller, R.A. and Macdonald, G.J. (1997b). Reply to comment on: Simultaneous presence of orbital inclination and eccentricity in proxy climate records from Ocean Drilling Program Site 806, by Schulz, M. and Mudelsee, M. *Geology* **25**: 861–2.

Munk, W., Dzieciuck, M. and Jayne, S. (2002). Millennial climate variability: is there a tidal connection? *J. Clim.* **15**: 370–85.

Munnecke, A., Westphal, H., Elrick, M. and Reijmer, J.J.G. (2001). The mineralogical composition of precursor sediments of calcareous rhythmites – a new approach. *Int. J. Earth Sci. (Geol. Rundsch.)* **90**: 795–812.

Murakoshi, N., Nakayama, N. and Masuda, F. (1995). Diurnal inequality pattern of tide in the upper Pleistocene Palaeo-Tokyo Bay: reconstruction from tidal deposits and growth-lines of fossil bivalves. *Spec. Pub. Int. Assoc. Sed.* **24**: 289–300.

Murray, B.C., Ward, W.R. and Yeung, S.C. (1973). Periodic insolation variation on Mars. *Science* **180**: 638–40.

Naish, T.R. and Kamp, P.J.J. (1997). Sequence stratigraphy of sixth order (41 k.y.) Pliocene-Pleistocene cyclothems, Wanganui Basin, New Zealand: a case for the regressive systems tract. *Geol. Soc. Am. Bull.* **109**: 978–99.

Naish, T.R., Abbott, S.T., Alloway, B.V., Beu, A.G., Carter, R.M., Edwards, A.R., Journeaux, T.J., Kamp, P.J.J., Pillans, B.J., Saul, G.S. and Woolfe, K.J. (1998). Astronomical calibration of a southern hemisphere Plio-Pleistocene reference section, Wanganui Basin, New Zealand. *Quat. Sci. Rev.* **17**: 695–710.

National Research Council (1994). *Solar Influences on Global Change*. National Academy Press, Washington, pp. 1–163.

Neff, U., Burns, S.J., Mangini, A., Mudelsee, M., Fleitmann, D. and Matter, A. (2001). Strong coherence between solar variability and the monsoon in Oman between 9 and 6 kyr ago. *Nature* **411**: 290–3.

Nicolis, C. and Nicolis, G. (1984). Is there a climatic attractor? *Nature* **311**: 529–32.

Ninnemann, U., Charles, C.D. and Hodell, D.A. (1999). Origin of global millennial scale climate events: constraints from the Southern Ocean deep sea sedimentary record. In: *Mechanisms of Global Climate Change at Millennial Time Scales*. Geophysical Monograph 112. American Geophysical Union, Washington, pp. 99–112.

Nowroozi, A.A. (1967). Table for Fisher's test of significance in harmonic analysis. *Geophys. J. R. Astron. Soc.* **12**: 512–20.

O'Brien, N.R. (1996). Shale lamination and sedimentary processes. In: *Palaeoclimatology and Palaeoceanography from Laminated Sediments*, Ed: A.E.S. Kemp. Geological Society Publication 116. The Geological Society, London, pp. 23–36.

Oeschger, H. and Beer, J. (1990). The past 5000 years history of solar modulation of cosmic radiation from ^{10}Be and ^{14}C

studies. *Philos. Trans. R. Soc. Lond.* **330A**: 471–80.

Olsen, G.H. (1977). *Modern Electronics Made Simple*. Allen, London, pp. 1–306.

Olsen, P.E. (1986). A 40 million year lake record of early Mesozoic orbital climatic forcing. *Science* **234**: 842–8.

Olsen, P.E. and Kent, D.V. (1996). Milankovitch climate forcing in the tropics of Pangaea during the Late Triassic. *Palaeogeog. Palaeoclimat. Palaeoecol.* **122**: 1–26.

Olsen, P.E. and Kent, D.V. (1999). Long-period Milankovitch cycles from the Late Triassic and Early Jurassic of eastern North America and their implications for the calibration of the Early Mesozoic time-scale and the long-term behaviour of the planets. *Philos. Trans. R. Soc. Lond.* **357**: 1761–86.

Oost, A.P., Dehaas, H., Ijnsen, F., Vandenboogert, J.M. and de Boer, P.L. (1993). The 18.6 yr nodal cycle and its impact on tidal sedimentation. *Sed. Geol.* **87**: 1–11.

Oppo, D.W. and Lehman, S.J. (1995). Suborbital timescale variability of North Atlantic deep water during the past 200,000 years. *Paleoceanography* **10**: 901–10.

Otnes, R.K. and Enochson, L. (1978). *Applied Time Series Analysis*, Volume 1. *Basic Techniques*. Wiley, Chichester, pp. 1–449.

Paillard, D., Labeyrie, L. and Yiou, P. (1996). Macintosh program performs time-series analysis. *EOS Trans. AGU* **77**: 379.

Pälike, H. and Shackleton, N.J. (2000). Constraints on astronomical parameters from the geological record for the past 25 Myr. *Earth Planet. Sci. Lett.* **182**: 1–14.

Pälike, H., Shackleton, N.J. and Röhl, U. (2001). Astronomical forcing in late Eocene marine sediments. *Earth Planet. Sci. Lett.* **193**: 589–602.

Pantev, C., Oostenveld, R., Engelien, A., Ross, B., Roberts, L.E. and Hoke, M. (1998). Increased auditory cortical representation in musicians. *Nature* **392**: 811–14.

Pardo-Igúzquiza, E. and Rodríguez-Tovar, F.J. (2000). The permutation test as a non-parametric method for statistical significance of power spectrum estimation in cyclostratigraphic research. *Earth Planet. Sci. Lett.* **181**: 175–89.

Pardo-Igúzquiza, E., Chica-Olmo, M. and Rodríguez-Tovar, F.J. (1994). CYSTRATI: a computer program for spectral analysis of stratigraphic successions. *Comput. Geosci.* **20**: 511–84.

Pardo-Igúzquiza, E., Schwarzacher, W. and Rodríguez-Tovar, F.J. (2000). A library of computer programs for assisting teaching and research in cyclostratigraphic analysis. *Comput. Geosci.* **26**: 723–40.

Park, J. and Herbert, T.D. (1987). Hunting for periodicities in a mid-Cretaceous sedimentary series. *J. Geophys. Res.* **92**: 14,027–40.

Paul, H.A., Zachos, J.C., Flower, B.P. and Tripati, A. (2000). Orbitally induced climate and geochemical variability across the Oligocene/Miocene boundary. *Paleoceanography* **15**: 471–85.

Pearce, R.B., Kemp, A.E.S., Baldauf, J.G. and King, S.C. (1995). High-resolution sedimentology and micropaleontology of laminated diatomaceous sediments from the eastern equatorial Pacific Ocean. *Proc. Ocean Drill. Prog. Sci. Res.* **138**: 647–63.

Pearn, W.C. (1964). Finding the ideal cyclothem. *Kansas Geol. Surv. Bull.* **169**: 399–413.

Pelletier, J.D. (1997). Analysis and modeling of the natural variability of climate. *J. Clim.* **10**: 1331–42.

Peper, T. and Cloetingh, S. (1995). Autocyclic perturbations of orbitally forced signals in the sedimentary record. *Geology* **23**: 937–40.

Percival, D.B. and Walden, A.T. (1993). *Spectral Analysis for Physical Applications. Multitaper and Conventional Univariate Techniques.* Cambridge University Press, Cambridge, pp. 1–583.

Pestiaux, P. and Berger, A. (1984a). An optimal approach to the spectral characteristics of deep sea climatic records. In: *Milankovitch and Climate*, Eds: A. Berger, J. Imbrie, J.D. Hays, G. Kukla, B. Saltzman, NATO ASI Series C126. Reidel, Dordrecht, volume 1, pp. 417–45.

Pestiaux, P. and Berger, A. (1984b). Impacts of deep-sea processes on paleoclimatic spectra. In: *Milankovitch and Climate*, Eds: A. Berger, J. Imbrie, J.D. Hays, G. Kukla, B. Saltzman, NATO ASI Series C126. Reidel, Dordrecht, volume 1, pp. 493–510.

Pestiaux, P., Duplessy, J.C. and Berger, A. (1987). Paleoclimatic variability at frequencies ranging from 10^{-4} cycle per year to 10^{-3} cycle per year – evidence for nonlinear behavior of the climate system. In: *Climate History, Periodicity and Predictability*, Eds: M. Rampino, J.E. Sanders, W.S. Newman and L.K., Kingsson. Van Nostrand Reinhold, New York, pp. 285–98.

Peters, S.E. and Foote, M. (2002). Determinants of extinction in the fossil record. *Nature* **416**: 420–4.

Petit, J.R., Jouzel, J., Raynaud, D., Barkov, N.I., Barnola, J.-M., Basile, I., Bender, M., Chappellez, J., Davis, M.,

Delaygue, G., Delmotte, M., Kotlyakov, V.M., Legrand, M., Lipenkov, V.Y., Lorius, C., Pépin, L., Ritz, C., Saltzman, E. and Stievenard, M. (1999). Climate and atmospheric history of the past 420,000 years from the Vostok ice core, Antarctica. *Nature* **399**: 429–36.

Petterson, G. (1996). Varved sediments in Sweden: a brief review. In: *Palaeoclimatology and Palaeoceanography from Laminated Sediments*, Ed: A.E.S. Kemp. Geological Society Publication 116. The Geological Society, London, pp. 73–7.

Philander, S.G. (1990). *El Niño, La Niña and the Southern Oscillation*. Academic Press, London, pp. 1–293.

Pilskaln, C.H. and Pike, J. (2001). Formation of Holocene sedimentary laminae in the Black Sea and the role of the benthic flocculent layer. *Paleoceanography* **16**: 1–19.

Pisias, N.G. (1983). Geologic time series from deep-sea sediments: time scales and distortion by bioturbation. *Mar. Geol.* **51**: 99–113.

Pisias, N.G. and Mix, A.C. (1988). Aliasing of the geologic record and the search for long-period Milankovitch cycles. *Paleoceanography* **3**: 613–19.

Pisias, N.G. and Moore, T.C. (1981). The evolution of Pleistocene climate: a time series approach. *Earth Planet. Sci. Lett.* **52**: 450–8.

Pisias, N.G., Mix, A.C. and Zahn, R. (1990). Nonlinear response in the global climate system: evidence from benthic oxygen isotope record in core RC13-110. *Paleoceanography* **5**: 147–60.

Plaut, G., Ghil, M. and Vautard, R. (1995). Interannual and interdecadal variability from a long temperature time series. *Science* **268**: 710–13.

Plotnick, R.E. (1986). A fractal model for the distribution of stratigraphic hiatuses. *J. Geol.* **94**: 885–90.

Prell, W.L., Imbrie, J., Martinson, D.G., Morley, J.J., Pisias, N.G., Shackleton, N.J. and Streeter, H.F. (1986). Graphic correlation of oxygen isotope stratigraphy: application to the Late Quaternary. *Paleoceanography* **1**: 137–62.

Press, W.H., Teukolsky, S.A., Vetterling, W.T. and Flannery, B.P. (1992). *Numerical Recipes, the Art of Scientific Computing*. Cambridge University Press, Cambridge, pp. 1–963.

Preston, F.W. and Henderson, J.H. (1964). Fourier series characterization of cyclic sediments for stratigraphic correlation. In: *Symposium on Cyclic Sedimentation*, Ed: D.F. Merriam. Bull Kansas Geol. Surv. **169**: 415–25.

Priestley, M.B. (1981). *Spectral Analysis and Time Series*. Academic Press, London, pp. 1–890.

Priestley, M.B. (1988). *Non-linear and Non-stationary Time Series Analysis*. Academic Press, London, pp. 1–237.

Proakis, J.G. and Menolakis, D.G. (1996). *Digital Signal Processing, Principles, Algorithms and Applications*. Prentice Hall, London, pp. 1–968.

Prokoph, A. and Agterberg, F.P. (1999). Detection of sedimentary cyclicity and stratigraphic completeness by wavelet analysis: an application of Late Albian cyclostratigraphy of the western Canada sedimentary basin. *J. Sed. Res.* **69**: 862–75.

Prokoph, A. and Barthelmes, F. (1996). Detection of nonstationarities in geological time series: wavelet transform of chaotic and cyclic sequences. *Comp. Geosci.* **22**: 1097–108.

Prokoph, A., Fowler, A.D. and Patterson, R.T. (2000). Evidence for periodicity and nonlinearity in a high-resolution fossil record of long-term evolution. *Geology* **28**: 867–70.

Pugh, D.T. (1987). *Tides, Surges and Mean Sea-level*. Wiley, Chichester, pp. 1–472.

Qin, X., Tan, M., Liu, T., Wang, X., Li, T. and Lu, J. (1999). Spectral analysis of 1000-year stalagmite lamina-thickness record from Shihua Cavern, Beijing, China, and its climatic significance. *The Holocene* **9**: 689–94.

Quinn, T.M., Taylor, F.W. and Crowley, T.J. (1993). A 173 year stable isotope record from a tropical south Pacific coral. *Quat. Sci. Rev.* **12**: 407–12.

Quinn, T.M., Crowley, T.J., Taylor, F.W., Henin, C., Joannot, P. and Join, Y. (1998). A multicentury stable isotope record from a New Caledonia coral: interannual and decadal sea surface temperature variability in the southwest Pacific since 1657 A.D. *Paleoceanography* **13**: 412–26.

Quinn, W.H. (1992). A study of Southern Oscillation-related climatic activity for AD622–1990 incorporating Nile River flood data. In: *El Niño: Historical and Paleoclimatic Aspects of the Southern Oscillation*, Eds: H.F. Diaz and V. Markgraf. Cambridge University Press, Cambridge, pp. 119–49.

Ramsay, A.T.S., Sykes, T.J.S. and Kidd, R.B. (1994a). Sedimentary hiatuses as indicators of fluctuating oceanic water masses: a new model. *J. Geol. Soc. Lond.* **151**: 737–40.

Ramsay, A.T.S., Sykes, T.J.S. and Kidd, R.B. (1994b). Waxing (and waning) lyrical on hiatuses: Eocene-Quaternary Indian Ocean hiatuses as proxy indicators of water mass production. *Paleoceanography* **9**: 857–77.

Räsänen, M.E., Linna, A.M., Santos, J.C.R. and Negri, F.R. (1995). Late Miocene tidal deposits in the Amazonian Foreland Basin. *Science* **269**: 386–9.

Rast, M.P., Fox, P.A., Lin, H., Lites, B.W., Meisner, R.W. and White, O.R. (1999). Bright rings around sunspots. *Nature* **401**: 678–9.

Raup, D.M. and Sepkoski, J.J. (1988). Testing for periodicity of extinction. *Science* **241**: 94–6.

Raymo, M.E., Ruddiman, W.F., Backman, J., Clement, B.M. and Martinson, D.G. (1989). Late Pliocene variations in northern hemisphere ice sheets and North Atlantic deep-water circulation. *Paleoceanography* **4**: 413–46.

Raymo, M.E., Ganley, K., Carter, S., Oppo, D.W. and McManus, J. (1998). Millennial-scale climate instability during the early Pleistocene epoch. *Nature* **392**: 699–702.

Reijmer, J.J.G., ten Kate, W.G.H.Z., Sprenger, A. and Schlager, W. (1991). Calciturbidite composition related to exposure and flooding of a carbonate platform (Triassic, Eastern Alps). *Sedimentology* **38**: 1059–74.

Rempel, A.W., Waddington, E.D., Wettlaufer, J.S. and Worster, M.G. (2001). Possible displacement of the climate signal in ancient ice by premelting and anomalous diffusion. *Nature* **411**: 568–71.

Rial, J.A. (1999). Pacemaking the ice ages by frequency modulation of Earth's orbital eccentricity. *Science* **285**: 564–8.

Richards, G.R. (1994). Orbital forcing and endogenous interactions: non-linearity, persistence and convergence in late Pleistocene climate. *Quat. Sci. Rev.* **13**: 709–25.

Ricken, W. (1986). *Diagenetic Bedding: A Model for Marl-Limestone Alternations*.

Lecture Notes Earth Science 6. Springer, Berlin, pp. 1–210.

Ricken, W. (1991a). Time span assessment – an overview. In: *Cycles and Events in Stratigraphy*, Eds: G. Einsele, W. Ricken and A. Seilacher. Springer, London, pp. 773–94.

Ricken, W. (1991b). Variation of sedimentation rates in rhythmically bedded sediments. Distinction between deposition types. In: *Cycles and Events in Stratigraphy*, Eds: G. Einsele, W. Ricken and A. Seilacher. Springer, London, pp. 167–87.

Ricken, W. (1993). *Sedimentation as a Three-Component System*. Springer, London.

Ricken, W. and Eder, W. (1991). Diagenetic modification of calcareous beds – an overview. In: *Cycles and Events in Stratigraphy*, Eds: G. Einsele, W. Ricken and A. Seilacher. Springer, London, pp. 430–49.

Ridgwell, A.J., Watson, A.J. and Raymo, M.E. (1999). Is the spectral signature of the 100 kyr glacial cycle consistent with a Milankovitch origin? *Paleoceanography* **14**: 437–40.

Riegel, W. (1991). Coal cyclothems and some models for their origin. In: *Cycles and Events in Stratigraphy*, Eds: G. Einsele, W. Ricken and A. Seilacher. Springer, London, pp. 733–50.

Ripepe, M. and Fischer, A.G. (1991). Stratigraphic rhythms synthesized from orbital variations. In: *Sedimentary Modeling: Computer Simulations and Methods for Improved Parameter Definition*, Eds: K. Franseen, W.L. Watney, C.G.St.C. Kendall and W. Ross. Kansas State Geol. Surv. Bull. **233**: 335–44.

Ripepe, M., Roberts, L.T. and Fischer, A.G. (1991). ENSO and sunspot cycles in varved Eocene oil shales from image analysis. *J. Sed. Petrol.* **61**: 1155–63.

Rittenour, T.M., Brigham-Grette, J. and Mann, M.E. (2000). El Niño-like climate teleconnections in New England during the Late Pleistocene. *Science* **288**: 1039–42.

Robock, A. (1996). Stratigraphic control of climate. *Science* **272**: 972–3.

Rodbell, D.T., Seltzer, G.O., Anderson, D.M., Abbott, M.B., Enfield, D.B. and Newman, J.H. (1999). An ∼15,000-year record of El Niño-driven alluviation in southwestern Ecuador. *Science* **283**: 516–20.

Rodwell, M.J., Rowell, D.P. and Folland, C.K. (1999). Oceanic forcing of the wintertime North Atlantic Oscillation and European climate. *Nature* **398**: 320–3.

Roulier, L.M. and Quinn, T.M. (1995). Seasonal- and decadal-scale climatic variability in southwest Florida during the middle Pliocene: inferences from a coralline stable isotope record. *Paleoceanography* **10**: 429–43.

Rubincam, D.P. (1995). Has climate changed the Earth's tilt? *Paleoceanography* **10**: 365–72.

Ruddiman, W.F. (1977). Late Quaternary deposition of ice-rafted sand in the subpolar North Atlantic (latitude 40°N to 65°N). *Geol. Soc. Am. Bull.* **88**: 1813–27.

Ruddiman, W.F. (1985). Climate studies in ocean cores. In: *Paleoclimate Analysis and Modelling*, Ed: A.D. Hecht. Kluwer, The Netherlands, pp. 197–257.

Ruddiman, W.F. and McIntyre, A. (1984). Ice-age thermal response and climatic role of the surface Atlantic Ocean, 40°N to 63°N. *Geol. Soc. Am. Bull.* **95**: 381–96.

Ruddiman, W.F., Raymo, M. and McIntyre, A. (1986). Matayama

41,000-year cycles, North Atlantic Ocean and northern hemisphere ice sheets. *Earth Planet. Sci. Lett.* **80**: 117–29.

Ruddiman, W.F., Raymo, M.E., Martinson, D.G., Clement, B.M. and Backman, J. (1989). Pleistocene evolution: northern hemisphere ice sheets and North Atlantic Ocean. *Paleoceanography* **4**: 353–412.

Rutherford, S. and D'Hondt, S. (2000). Early onset and tropical forcing of 100,000-year Pleistocene glacial cycles. *Nature* **408**: 72–5.

Sadler, P.M. (1981). Sediment accumulation rates and the completeness of stratigraphic sections. *J. Geol.* **89**: 569–84.

Sadler, P.M. and Strauss, D.J. (1990). Estimation of completeness of stratigraphical sections using empirical data and theoretical models. *J. Geol. Soc. Lond.* **147**: 471–85.

Saji, N.H., Goswami, B.N., Vinayachandran, P.N. and Yamagata, T. (1999). A dipole mode in the tropical Indian Ocean. *Nature* **401**: 360–3.

Saltzman, B. and Verbitsky, M. (1994). Late Pleistocene climatic trajectory in the phase space of global ice, ocean state, and CO_2: observations and theory. *Paleoceanography* **9**: 767–79.

Sander, B. (1936). Beiträge zur Kenntnis der Anlargerungsgefüge. *Mineral. Petrogr. Mitt.* **48**: 27–139.

Schaaf, M. and Thurow, J. (1997). Tracing short cycles in long records: the study of inter-annual to inter-centennial climate change from long sediment records, examples from the Santa Barbara Basin. *J. Geol. Soc. Lond.* **154**: 613–22.

Schiffelbein, P. (1984). Effect of benthic mixing on the information content of deep sea stratigraphical signals. *Nature* **311**: 651–3.

Schiffelbein, P. and Dorman, L. (1986). Spectral effects of time-depth nonlinearities in deep sea sediment records: a demodulation technique for realigning time and depth scales. *J. Geophys. Res.* **91**: 3821–35.

Schlesinger, M.E. and Ramankutty, N. (1994). An oscillation in the global climate system of period 65–70 years. *Nature* **367**: 723–6.

Schmitz, W.J. and McCarthey, M.S. (1993). On the North Atlantic circulation. *Rev. Geophys.* **31**: 29–49.

Scholz, C.H. (1998). Earthquakes and friction laws. *Nature* **391**: 37–42.

Schulz, H., von Rad, U. and Erlenkeuser, H. (1998). Correlation between Arabian Sea and Greenland climate oscillations of the past 110,000 years. *Nature* **393**: 54–7.

Schulz, M. and Stattegger, K. (1997). Spectrum: spectral analysis of unevenly spaced paleoclimatic time series. *Comput. Geosci.* **23**: 929–45.

Schuster, A. (1898). On the investigation of hidden periodicities with application to a supposed 26-day period of meteorological phenomenon. *Terr. Mag. Atmos. Elect.* **3**: 13–41.

Schwarzacher, W. (1964). An application of statistical time-series analysis of a limestone-shale sequence. *J. Geol.* **72**: 195–213.

Schwarzacher, W. (1975). *Sedimentation Models and Quantitative Stratigraphy.* Elsevier, Amsterdam, pp. 1–382.

Schwarzacher, W. (1991). Milankovitch cycles and the measurement of time. In: *Cycles and Events in Stratigraphy*, Eds: G. Einsele, W. Ricken, and A. Seilacher. Springer, London, pp. 855–63.

Schwarzacher, W. (1993). *Cyclostratigraphy and the Milankovitch Theory.* Elsevier, Amsterdam, pp. 1–225.

Schwarzacher, W. (1998). Stratigraphic resolution, cycles and sequences. In: *Sequence Stratigraphy – Concepts and Applications*, Eds: F.M. Gradstein, K.O. Sandvik and N.J. Milton. Norwegian Petroleum Society Special Publication No. 8. Elsevier, Amsterdam, pp. 1–8.

Schwarzacher, W. and Fischer, A.G. (1982). Limestone-shale bedding and perturbations of the Earth's orbit. In: *Cyclic and Event Stratification*, Eds: G. Einsele and A. Seilacher. Springer, London, pp. 72–95.

Schweingruber, F.H., Echstein, D., Serre-Bachet, F. and Braeker, O.U. (1990). Identification, presentation and interpretation of event years and pointer years in dendrochronology. *Dendrochronologia* **8**: 9–38.

Seidov, D. and Maslin, M. (1999). North Atlantic deep water circulation collapse during Heinrich events. *Geology* **27**: 23–6.

Sellwood, B.W. (1970). The relation of trace fossils to small scale sedimentary cycles in the British Lias. In: *Trace Fossils*, Eds: T.P. Crimes and J.C. Harper. Seel House Press, Liverpool, pp. 489–504.

Sepkoski, J.J. (1989). Periodicity in extinction and the problem of catastrophism in the history of life. *J. Geol. Soc.* **146**: 7–19.

Sepkoski, J.J. and Raup, D.M. (1986). Periodicity in marine extinction events. In*: Dynamics of Extinction*, Ed: D.K. Elliott. Wiley, Chichester, pp. 3–36.

Shackleton, N.J. (2000). The 100,000-year ice-age cycle identified and found to lag temperature, carbon dioxide, and orbital eccentricity. *Science* **289**: 1897–902.

Shackleton, N.J. and Crowhurst, S. (1997). Sediment fluxes based on an orbitally tuned time scale 5 Ma to 14 Ma, Site 926. *Proc. Ocean Drill. Prog.* **154**: 69–82.

Shackleton, N.J. and Imbrie, J. (1990). The $\delta^{18}O$ spectrum of oceanic deep water over a five-decade band. *Clim. Change* **16**: 217–30.

Shackleton, N.J. and Opdyke, N.D. (1973). Oxygen isotope and palaeomagnetic stratigraphy of equatorial Pacific core V28-238: oxygen isotope temperatures and ice volumes on a 10^5 year and 10^6 year scale. *Quat. Res.* **3**: 39–55.

Shackleton, N.J. and Pisias, N.G. (1985). Atmospheric carbon dioxide, orbital forcing and climate. In: *The Carbon Cycle and Atmospheric CO$_2$: Natural Variations Archaean to Present*, Eds: E.T. Sundquist and W.S. Broecker. Geophysical Monograph 32. American Geophysical Union, Washington, pp. 303–17.

Shackleton, N.J., Berger, A. and Peltier, W.R. (1990). An alternative astronomical calibration of the lower Pleistocene timescale based on ODP Site 677. *Trans. R. Soc. Edinburgh, Earth Sci.* **81**: 251–61.

Shackleton, N.J., Crowhurst, S.J., Hagelberg, T., Pisias, N.G. and Schneider, D.A. (1995a). A new late Neogene timescale: application to Leg 138 sites. *Proc. Ocean Drill. Prog. Sci. Res.* **138**: 73–101.

Shackleton, N.J., Hagelberg, T.K. and Crowhurst, S.J. (1995b). Evaluating the success of astronomical tuning: pitfalls of using coherence as a criterion for assessing pre-Pleistocene timescales. *Paleoceanography* **10**: 693–7.

Shackleton, N.J., Hall, M.A. and Pate, D. (1995c). Pliocene isotope stratigraphy of Site 846. *Proc. Ocean Drill. Prog. Sci. Res.* **138**: 337–55.

Shackleton, N.J., Crowhurst, S.J., Weedon, G.P. and Laskar, J. (1999a). Astronomical calibration of

Oligocene-Miocene time. *Philos. Trans. R. Soc. Lond.* **357**: 1907–29.

Shackleton, N.J., McCave, I.N. and Weedon, G.P. (Eds) (1999b). Astronomical (Milankovitch) calibration of the geological time-scale. *Philos. Trans. R. Soc. Lond.* **357**: 1733–2007.

Shen, G.T., Cole, J.E., Lea, D.W., Linn, L.J., McConnaughey, E.A. and Fairbanks, R.G. (1992). Surface ocean variability at Galapagos from 1936–1982: calibration of geochemical tracers in corals. *Paleoceanography* **7**: 563–88.

Shindell, D., Rind, D., Balachandran, N., Lean, J. and Lonergan, P. (1999). Solar cycle variability, ozone and climate. *Science* **284**: 305–8.

Shindell, D.T., Schmidt, G.A., Mann, M.E., Rind, D. and Waple, A. (2001). Solar forcing of regional climate during the Maunder Minimum. *Science* **294**: 2149–52.

Short, D.A., Mengel, J.G., Crowley, T.J., Hyde, W.T. and North, G.R. (1991). Filtering of Milankovitch cycles by Earth's geography. *Quat. Res.* **35**: 157–73.

Sigman, D.M. and Boyle, E.A. (2000). Glacial/interglacial variations in atmospheric carbon dioxide. *Nature* **407**: 859–69.

Sloan, L.C. and Huber, M. (2001). Eocene oceanic responses to orbital forcing on precessional time scales. *Paleoceanography* **16**: 101–11.

Smith, D.G. (1994). Cyclicity or chaos? Orbital forcing versus non-linear dynamics. In: *Orbital Forcing and Cyclic Sequences*, Eds: P.L. de Boer and D.G. Smith. International Association of Sedimentologists Special Publication 19. Blackwell, Oxford, pp. 531–44.

Smith, N.D., Phillips, A.C. and Powell, R.D. (1990). Tidal drawdown: a mechanism for producing cyclic sediment laminations in glaciomarine deltas. *Geology* **18**: 10–13.

Solanki, S.K., Schüssler, M. and Fligge, M. (2000). Evolution of the Sun's large-scale magnetic field since the Maunder minimum. *Nature* **408**: 445–7.

Sonett, C.P. and Chan, M.A. (1998). Neoproterozoic Earth–Moon dynamics: rework of the 900Ma Big Cottonwood Canyon tidal laminae. *Geophys. Res. Lett.* **25**: 539–42.

Sonett, C.P. and Finney, S.A. (1990). The spectrum of radiocarbon. *Philos. Trans. R. Soc. Lond.* **330A**: 413–26.

Sonett, C.P. and Williams, G.E. (1985). Solar periodicities expressed in varves from glacial Skilak Lake, southern Alaska. *J. Geophys. Res.* **90**: 12,019–26.

Sonett, C.P., Finney, S.A. and Williams, C.P. (1988). The lunar orbit in the late Precambrian and the Elatina sandstone laminae. *Nature* **335**: 806–8.

Sonett, C.P., Kvale, E.P., Zakharian, A., Chan, M.J. and Demko, T.M. (1996). Late Proterozoic and Paleozoic tides, retreat of the Moon, and rotation of the Earth. *Science* **273**: 100–4.

Spencer-Cervato, C. (1998). Changing depth distribution of hiatuses during the Cenozoic. *Paleoceanography* **13**: 178–82.

Stauffer, B., Blunier, T., Dällenbach, A., Indermühle, A., Scwander, J., Stocker, T.F., Tschumi, J., Chappellaz, J., Raynaud, D., Hammer, C.U. and Clausen, H.B. (1998). Atmospheric CO_2 concentration and millennial-scale climate change during the last glacial period. *Nature* **392**: 59–62.

Stewart, I. (1990). *Does God Play Dice? The New Mathematics of Chaos.* Penguin, London, pp. 1–317.

Stigler, S.M. and Wagner, M.J. (1987). A substantial bias in nonparametric tests for periodicity in geophysical data. *Science* **238**: 940–5.

Stigler, S.M. and Wagner, M.J. (1988). Testing for periodicity of extinction. *Science* **241**: 96–9.

Stockton, C.W., Boggess, W.R. and Meko, D.M. (1985). Climate and tree rings. In: *Paleoclimate Analysis and Modelling*, Ed: A.D. Hecht. Wiley, Chichester, pp. 71–161.

Strauss, D. and Sadler, P.M. (1989). Stochastic models for the completeness of stratigraphic sections. *J. Int. Assoc. Math. Geol.* **21**: 37–59.

Street-Perrott, F. and Perrott, R.A. (1990). Abrupt fluctuations in the tropics: the influence of Atlantic Ocean circulation. *Nature* **343**: 607–12.

Stuiver, M. and Braziunas, T.F. (1989). Atmospheric ^{14}C and century-scale solar oscillations. *Nature* **338**: 405–7.

Stuiver, M. and Quay, P.D. (1980). Changes in atmospheric carbon-14 attributed to a variable sun. *Science* **207**: 11–19.

Stuiver, M., Grootes, P.M. and Braziunas, T.F. (1995). The GISP2 $\delta^{18}O$ climate record of the past 16,500 years and the role of the Sun, oceans and volcanoes. *Quat. Res.* **44**: 341–54.

Stuiver, M., Braziunas, T.F., Grootes, P.M. and Zielinski, G.A. (1997). Is there evidence for solar forcing of climate in the GISP2 oxygen isotope record? *Quat. Res.* **48**: 259–66.

Sugihara, G. and May, R.M. (1990). Nonlinear forecasting as a way of distinguishing chaos from measurement error in time series. *Nature* **344**: 734–41.

Sutton, R.T. and Allen, M.R. (1997). Decadal predictability of North Atlantic sea surface temperature and climate. *Nature* **388**: 563–7.

Sztanó, O. and de Boer, P.L. (1995). Basin dimensions and morphology as controls on amplification of tidal motions (the Early Miocene North Hungarian Bay). *Sedimentology* **42**: 665–82.

Taner, M.T., Koehler, F. and Sheriff, R.E. (1979). Complex seismic trace analysis. *Geophysics* **44**: 1041–63.

Taylor, A.H., Jordan, M.B. and Stephens, J.A. (1998). Gulf Stream shifts following ENSO events. *Nature* **393**: 638.

Taylor, C.A. (1965). *Physics of Musical Sounds*. English University Press, Aylesbury, pp. 1–196.

Taylor, C.A. (1976). *Sounds of Music*. BBC, London, pp. 1–183.

Taylor, K.C., Lamorey, G.W., Doyle, G.A., Alley, R.B., Grootes, P.M., Mayewski, P.A., White, J.W.C. and Barlow, L.K. (1993). The "flickering switch" of late Pleistocene climate change. *Nature* **361**: 432–6.

Tessier, B. and Gigot, P. (1989). A vertical record of different tidal cyclicities: an example from the Miocene marine molasse of Digne (Haute Provence, France). *Sedimentology* **36**: 767–76.

Thomson, D.J. (1982). Spectrum estimation and harmonic analysis. *Proc. IEEE* **70**: 1055–96.

Thomson, D.J. (1990). Quadratic-inverse spectrum estimates; applications to paleoclimatology. *Philos. Trans. R. Soc. Lond.* **332A**: 539–97.

Thomson, D.J. (1995). The seasons, global temperature, and precession. *Science* **268**: 59–68.

Thomson, J., Higgs, N.C. and Clayton, T. (1995). A geochemical criterion for the recognition of Heinrich events and estimation of their depositional fluxes by the $^{230}Th_{excess}$ profiling method. *Earth Planet. Sci. Lett.* **135**: 29–43.

Thunell, R., Pride, C., Tappa, E. and
Muller-Karger, F. (1993). Varve
formation in the Gulf of California:
insights from time series sediment trap
sampling and remote sensing. *Quat. Sci.
Rev.* **12**: 451–64.

Thunell, R.C., Tappa, E. and Anderson,
D.M. (1995). Sediment fluxes and
varve formation in Santa Barbara
Basin, offshore California. *Geology* **23**:
1083–6.

Tiedemann, R. and Franz, S.O. (1997).
Deep-water circulation, chemistry,
and terrigenous sediment supply in
the equatorial Atlantic during the
Pliocene, 3.3–2.6 Ma and 5–4.5 Ma.
Proc. Ocean Drill. Prog. Sci. Res. **154**:
299–318.

Tinsley, B.A. (1996). Correlations of
atmospheric dynamics with solar wind
induced changes of air-Earth current
density into cloud tops. *J. Geophys. Res.*
101: 29,701–14.

Tinsley, B.A. (1997). Do effects of global
atmospheric electricity on clouds cause
climate changes? *EOS Trans. Am.
Geophys. Union* **78**: 341–9.

Tipper, J.C. (1983). Rates of sedimentation,
and stratigraphical completeness. *Nature*
302: 696–8.

Tong, H. (1990). *Non-linear Time Series.*
Oxford University Press, Oxford,
pp. 1–564.

Toon, O.B., Pollack, J.B., Ward, W.,
Burns, J.A. and Bilski, K. (1980). The
astronomical theory of climate change on
Mars. *Icarus* **44**: 552–607.

Torrence, C. and Compo, G.P. (1998). A
practical guide to wavelet analysis.
Bull. Am. Meteorol. Soc. **79**:
61–78.

Trauth, M.H. (1998). TURBO: a
dynamic-probabilistic simulation to study
the effects of bioturbation on

paleoceanographic time series. *Comput.
Geosci.* **24**: 433–41.

Trauth, M.H., Sarnthein, M. and Arnold, M.
(1997). Bioturbational mixing depth and
carbon flux at the seafloor.
Paleoceanography **12**: 517–26.

Tsonis, A.A. and Elsner, J.B. (1992).
Nonlinear prediction as a way of
distinguishing chaos from random fractal
sequences. *Nature* **358**: 217–20.

Tudhope, A.W., Shimmield, G.B.,
Chilcott, C.P., Jebb, M., Fallick, A.E. and
Dalgleish, A.N. (1995). Recent changes
in climate in the far western equatorial
Pacific and their relationship to the
Southern Oscillation; oxygen isotope
records from massive corals, Papua New
Guinea. *Earth Planet. Sci. Lett.* **136**:
575–90.

Turcotte, D.L. (1997). *Fractals and Chaos
in Geology and Geophysics*, 2nd Edition.
Cambridge University Press, Cambridge,
pp. 1–398.

Tziperman, E., Stone, L., Cane, M.A. and
Jarosh, H. (1994). El Niño chaos:
overlapping of resonances between the
seasonal cycle and the Pacific
ocean-atmosphere oscillator. *Science* **264**:
72–4.

Ulrych, T.J. and Bishop, T.N. (1975).
Maximum entropy spectral analysis and
autoregressive decomposition. *Rev.
Geophys. Space Phys.* **13**: 183–200.

Urban, F.E., Cole, J.E. and Overpeck, J.T.
(2000). Influence of mean climate change
on climate variability from a 155-year
tropical Pacific coral record. *Nature* **407**:
989–93.

Vail, P.R., Mitchum, R.M., Todd, R.G.,
Widmer, J.W., Thompson, S.,
Sangree, J.B., Bubb, J.N. and Hatlelid,
W.G. (1977). Seismic stratigraphy and
global changes of sealevel. In: *Seismic
Stratigraphy – Application to*

Hydrocarbon Exploration, Ed: C.E. Payton. Am. Assoc. Petrol. Geol. Mem. **26**: 46–212.

Vail, P.R., Audemard, F., Bowman, S.A., Eisner, P.N. and Perez-Cruz, C. (1991). The stratigraphic signatures of tectonics, eustasy and sedimentology – an overview. In: *Cycles and Events in Stratigraphy*, Eds: G. Einsele, W. Ricken and A. Seilacher. Springer, London, pp. 617–59.

Van Echelpoel, E. (1994). Identification of regular sedimentary cycles using Walsh spectral analysis with results from the Boom Clay Formation, Belgium. In: *Orbital Forcing and Cyclic Sequences*, Eds: P.L. de Boer, and D.G. Smith. International Association of Sedimentologists Special Publication 19. Blackwell, Oxford, pp. 63–74.

van Geel, B., Raspopov, O.M., Renssen, H., van der Plicht, J., Dergachev, V.A. and Meijer, H.A.J. (1999). The role of solar forcing upon climate change. *Quat. Sci. Rev.* **18**: 331–8.

Vautard, R. and Ghil., M. (1989). Singular spectrum analysis in non-linear dynamics, with applications to palaeoclimatic time series. *Physica D* **35**: 395–424.

Visser, M.J. (1980). Neap-Spring cycles reflected in Holocene subtidal large-scale bedform deposits: a preliminary note. *Geology* **8**: 543–6.

Wales, D.J. (1991). Calculating the rate of loss of information from chaotic time series by forecasting. *Nature* **350**: 485–8.

Walther, G. (1997). Absence of correlation between the solar neutrino flux and the sunspot number. *Phys. Rev. Lett.* **79**: 4522–4.

Wanless, H.R. and Weller, J.M. (1932). Correlation and extent of Pennsylvanian cyclothems. *Geol. Soc. Am. Bull.* **43**: 1003–16.

Webster, P.J. and Palmer, T.N. (1997). The past and the future of El Niño. *Nature* **390**: 562–4.

Webster, P.J., Moore, A.M., Loschnigg, J.P. and Leben, R.R. (1999). Coupled ocean-atmosphere dynamics in the Indian Ocean during 1997–98. *Nature* **401**: 356–60.

Weedon, G.P. (1989). The detection and illustration of regular sedimentary cycles using Walsh power spectra and filtering, with examples from the Lias of Switzerland. *J. Geol. Soc. Lond.* **146**: 133–44.

Weedon, G.P. (1993). The recognition and stratigraphic implications of orbital forcing of climate and sedimentary cycles. In: *Sedimentology Review*, Ed: V.P. Wright. Blackwell, Oxford, pp. 31–50.

Weedon, G.P. and Jenkyns, H.C. (1990). Regular and irregular climatic cycles and the Belemnite Marls (Pliensbachian, Lower Jurassic, Wessex Basin). *J. Geol. Soc. Lond.* **147**: 915–18.

Weedon, G.P. and Jenkyns, H.C. (1999). Cyclostratigraphy and the Early Jurassic timescale: data from the Belemnite Marls, Dorset, southern England. *Geol. Soc. Am. Bull.* **111**: 1823–40.

Weedon, G.P. and Read, W.A. (1995). Orbital-climatic forcing of Namurian cyclic sedimentation from spectral analysis of the Limestone Coal Group, Central Scotland. In: *Orbital Forcing Timescales and Cyclostratigraphy*, Eds: M.R. House and A.S. Gale. Geological Society Special Publication No. 85. The Geological Society, London, pp. 51–66.

Weedon, G.P. and Shimmield, G.B. (1991). Late Pleistocene upwelling and productivity variations in the northwest Indian Ocean deduced from spectral analyses of geochemical data from sites 722 and 724. *Proc. Ocean Drill. Program Sci. Res.* **117**: 431–43.

Weedon, G.P., Shackleton, N.J. and Pearson, P.N. (1997). The Oligocene timescale and cyclostratigraphy on the Ceara Rise, western equatorial Atlantic. *Proc. Ocean Drill. Prog. Sci. Res.* **154**: 101–14.

Weedon, G.P., Jenkyns, H.C., Coe, A.L. and Hesselbo, S.P. (1999). Astronomical calibration of the Jurassic time-scale from cyclostratigraphy in British mudrock formations. *Philos. Trans. R. Soc. Lond.* **357**: 1787–813.

Weiss, N.O. (1990). Periodicity and aperiodicity in solar magnetic activity. *Philos. Trans. R. Soc. Lond.* **330A**: 617–25.

Wells, J.W. (1963). Coral growth and geochronometry. *Nature* **197**: 948–50.

Weltje, G. and de Boer, P.L. (1993). Astronomically induced paleoclimatic oscillations reflected in Pliocene turbidite deposits on Corfu (Greece): implications for the interpretation of higher order cyclicity in ancient turbidite systems. *Geology* **21**: 307–10.

Whitcombe, L.J. (1996). A FORTRAN program to calculate tidal heights using the simplified harmonic method of tidal prediction. *Comp. Geosci.* **22**: 817–21.

White, J.C.W., Barlow, L.K., Fisher, D., Grootes, P., Jouzel, J., Johnsen, S.J., Stuiver, M. and Clausen, H. (1997). The climate signal in the stable isotopes of snow from Summit, Greenland: results of comparisons with modern climate observations. *J. Geophys. Res.* **102**: 26,425–39.

Wiesenfield, K. and Moss, F. (1995). Stochastic resonance and the benefits of noise: from ice ages to crayfish and SQUIDS. *Nature* **373**: 33–6.

Wilkinson, B.H., Drummond, C.N., Rothman, E.D. and Diedrich, N.W.

(1997). Stratal order in peritidal carbonate sequences. *J. Sed. Res.* **67**: 1068–82.

Wilkinson, B.H., Drummond, C.N., Diedrich, N.W. and Rothman, E.D. (1999). Poisson processes of carbonate accumulation on Paleozoic and Holocene platforms. *J. Sed. Res.* **69**: 338–50.

Williams, G.E. (1981). Sunspot periods in the late Precambrian glacial climate and solar-planetary relations. *Nature* **291**: 624–8.

Williams, G.E. (1986). The solar cycle in Precambrian time. *Sci. Am.* **255**: 88–95.

Williams, G.E. (1988). Cyclicity in the Late Precambrian Elatina Formation, South Australia: solar or tidal signature? *Clim. Change* **13**: 117–28.

Williams, G.E. (1989). Late Precambrian tidal rhythmites in South Australia and the history of the Earth's rotation. *J. Geol. Soc. Lond.* **146**: 97–111.

Williams, G.E. (1991). Milankovitch-band cyclicity in bedded halite deposits contemporaneous with Late Ordovician-Early Silurian glaciation, Canning Basin, western Australia. *Earth Planet. Sci. Lett.* **103**: 143–55.

Williams, G.E. and Sonett, C.P. (1985). Solar signature in sedimentary cycles from the late Precambrian Elatina Formation, Australia. *Nature* **318**: 523–7.

Williams, G.P. (1997). *Chaos Theory Tamed*. National Academy Press, Washington, pp. 1–499.

Williams, R.B.G. (1984). *Introduction to Statistics for Geographers and Earth Scientists*. Macmillan, London, pp. 1–349.

Willson, R.C. and Hudson, H.S. (1988). Solar luminosity variations in solar cycle 21. *Nature* **332**: 810–12.

Willson, R.C. and Hudson, H.S. (1991). The Sun's luminosity over a complete solar cycle. *Nature* **351**: 42–4.

Wilson, D.S. (1993). Confirmation of the astronomical calibration of the magnetic polarity timescale from sea-floor spreading rates. *Nature* **364**: 788–90.

Worthington, P.F. (1990). Sediment cyclicity from well logs. In: *Geological Applications of Wireline Logs*, Eds: A. Hurst, M.A. Lovell and A.C. Morton. Geological Society Special Publication No. 48. The Geological Society, London, pp. 123–32.

Wunsch, C. (1999). The interpretation of short climate records, with comments on the North Atlantic and Southern Oscillations. *Bull. Am. Meteorol. Soc.* **80**: 245–55.

Wunsch, C. (2000a). On sharp spectral lines in the climate record and the millennial peak. *Paleoceanography* **15**: 417–24.

Wunsch, C. (2000b). Moon, tides and climate. *Nature* **405**: 743–4.

Yang, C-S. and Baumfalk, Y.A. (1994). Milankovitch cyclicity in the Upper Rotliegend Group of the Netherlands offshore. In: *Orbital Forcing and Cyclic Sequences*, Eds: P.L. de Boer and D.G. Smith. International Association of Sedimentologists Special Publication No. 19. Blackwell, Oxford, pp. 47–61.

Yang, C-S. and Nio, S-D. (1985). The estimation of palaeohydrodynamic processes from subtidal deposits using time series analysis methods. *Sedimentology* **32**: 41–57.

Yiou, P., Baert, E. and Loutre, M.F. (1996). Spectral analysis of climate data. *Surve. Geophys.* **17**: 619–63.

Yiou, P., Sornette, D. and Ghil, M. (2000). Data-adaptive wavelets and multi-scale singular-spectrum analysis. *Physica D* **142**: 254–90.

Young, P.C. (1999). Nonstationary time series analysis. *Prog. Environ. Sci.* **1**: 3–48.

Yu, Z. and Ito, E. (1999). Possible forcing of century-scale drought frequency in the northern Great Plains. *Geology* **27**: 263–6.

Zachos, J.C., Flower, B.P. and Paul, H. (1997). Orbitally paced climate oscillations across the Oligocene/Miocene boundary. *Nature* **388**: 567–70.

Zachos, J.C., Shackleton, N.J., Revenaugh, J.S., Palike, H. and Flower, B.P. (2001). Climate response to orbital forcing across the Oligocene-Miocene boundary. *Science* **292**: 274–8.

Zhang, R-H., Rothstein, L.M. and Busalacchi, A.J. (1998). Origin of upper-ocean warming and El Niño change on decadal scales in the tropical Pacific Ocean. *Nature* **391**: 879–83.

Zolitschka, B. (1996). Image analysis and microscopic investigation of annually laminated lake sediments from Fatetteville Green Lake (NY, USA) Lake C2 (NWT, Canada) and Holzmaar (Germany): a comparison. In: *Palaeoclimatology and Palaeoceanography from Laminated Sediments*, Ed: A.E.S. Kemp. Geological Society Special Publication No. 116. The Geological Society, London, pp. 49–55.

Index